Environmental Restoration of
Metal-Contaminated Soils

Environmental Restoration of Metal-Contaminated Soils

Editors

Fulvia Chiampo
Massimo Zacchini

MDPI • Basel • Beijing • Wuhan • Barcelona • Belgrade • Manchester • Tokyo • Cluj • Tianjin

Editors
Fulvia Chiampo
Department of Applied Science
and Technology, Politecnico di
Torino
Italy

Massimo Zacchini
Research Institute on Terrestrial
Ecosystem, National Research
Council of Italy
Italy

Editorial Office
MDPI
St. Alban-Anlage 66
4052 Basel, Switzerland

This is a reprint of articles from the Special Issue published online in the open access journal *Applied Sciences* (ISSN 2076-3417) (available at: http://www.mdpi.com).

For citation purposes, cite each article independently as indicated on the article page online and as indicated below:

LastName, A.A.; LastName, B.B.; LastName, C.C. Article Title. *Journal Name* **Year**, *Volume Number*, Page Range.

ISBN 978-3-0365-3363-6 (Hbk)
ISBN 978-3-0365-3364-3 (PDF)

© 2022 by the authors. Articles in this book are Open Access and distributed under the Creative Commons Attribution (CC BY) license, which allows users to download, copy and build upon published articles, as long as the author and publisher are properly credited, which ensures maximum dissemination and a wider impact of our publications.

The book as a whole is distributed by MDPI under the terms and conditions of the Creative Commons license CC BY-NC-ND.

Contents

About the Editors vii

Preface to "Environmental Restoration of Metal-Contaminated Soils" ix

Fulvia Chiampo and Massimo Zacchini
Environmental Restoration of Metal-Contaminated Soils
Reprinted from: *Appl. Sci.* 2021, *11*, 10805, doi:10.3390/app112210805 1

Carla Maria Raffa, Fulvia Chiampo and Subramanian Shanthakumar
Remediation of Metal/Metalloid-Polluted Soils: A Short Review
Reprinted from: *Appl. Sci.* 2021, *11*, 4134, doi:10.3390/app11094134 5

Daniela Zingaretti and Renato Baciocchi
Different Approaches for Incorporating Bioaccessibility of Inorganics in Human Health Risk Assessment of Contaminated Soils
Reprinted from: *Appl. Sci.* 2021, *11*, 3005, doi:10.3390/app11073005 29

Claudio Cameselle, Susana Gouveia and Adrian Cabo
Enhanced Electrokinetic Remediation for the Removal of Heavy Metals from Contaminated Soils
Reprinted from: *Appl. Sci.* 2021, *11*, 1799, doi:10.3390/app11041799 43

Vera Yurak, Rafail Apakashev, Alexey Dushin, Albert Usmanov, Maxim Lebzin and Alexander Malyshev
Testing of Natural Sorbents for the Assessment of Heavy Metal Ions' Adsorption
Reprinted from: *Appl. Sci.* 2021, *11*, 3723, doi:10.3390/app11083723 55

Elnaz Amirahmadi, Seyed Mohammad Hojjati, Claudia Kammann, Mohammad Ghorbani and Pourya Biparva
The Potential Effectiveness of Biochar Application to Reduce Soil Cd Bioavailability and Encourage Oak Seedling Growth
Reprinted from: *Appl. Sci.* 2020, *10*, 3410, doi:10.3390/app10103410 65

Elisa Gaggero, Paola Calza, Debora Fabbri, Anna Fusconi, Marco Mucciarelli, Giorgio Bordiglia, Ornella Abollino and Mery Malandrino
Assessment and Mitigation of Heavy Metals Uptake by Edible Vegetables Grown in a Turin Contaminated Soil Used as Vegetable Garden
Reprinted from: *Appl. Sci.* 2020, *10*, 4483, doi:10.3390/app10134483 79

Jung-Geun Han, Dongho Jung, Jong-Young Lee, Dongchan Kim and Gigwon Hong
A Study on the Flow Characteristics of Copper Heavy Metal Microfluidics with Hydrophobic Coating and pH Change
Reprinted from: *Appl. Sci.* 2021, *11*, 4328, doi:10.3390/app11104328 95

About the Editors

Fulvia Chiampo, Prof., is currently based at the Department of Applied Science and Technology of Politecnico di Torino. She is a professor of Chemical Equipment Design. At present, her scientific activity is mainly focused on environmental issues, with a particular emphasis on solid waste treatment, landfill biogas, soil remediation, and industrial wastewater treatment. In the past, as a chemical engineer, she carried out research activities in the chemical reactor and food engineering sectors. Her expertise on environmental topics is applied to support local authorities and municipalities involved in environmental projects, especially solid wastes. She is the author and co-author of more than 80 scientific papers, published in international journals or presented at international conferences. At an international level, Prof. Chiampo has coordinated projects under the programs TEMPUS IV, Life+ 2009, and the Executive Programme of Scientific and Technological Cooperation between the Republic of India and the Italian Republic (2017–2019).

Massimo Zacchini, Ph.D., is currently working at the Research Institute on Terrestrial Ecosystems (IRET) of CNR. His main research interests concern the physiological and biochemical characterization of plants useful for the decontamination of polluted soils and water (phytoremediation) and the ecotoxicological risk assessment of chemicals in plants. He is the author and co-author of more than 60 scientific papers, mainly published in ISI journals. He has been involved in numerous research projects funded by the EU and Italian and Spanish Ministry of Science and Research, and CNR Joint projects with foreign research institutions. He has organized international conferences and workshops dealing with innovative technologies for the decontamination of soil and water, also acting as an invited speaker and chairperson in many international conferences on bioremediation and environmental pollution. He has carried out academic activities as a lecturer and member of the evaluation panels of European and national Ph.D. students, and tutoring activities for Ph.D. and Erasmus+ students, as well as European and national research fellows. He is a member of the Editorial Board of international scientific journals and a referee for many ISI journals dealing with phytoremediation, plant physiology, plant biochemistry, and environmental science.

Preface to "Environmental Restoration of Metal-Contaminated Soils"

This Special Issue entitled "Environmental Restoration of Metal-Contaminated Soils" focuses on the issues linked to soils contaminated with heavy metals and metalloids, dealing with current research activities around the world at the laboratory and field scale. These activities are the pillars for the application of strategies on a real-world scale, to remediate industrial soils affected by the problem. When an industrial soil contains pollutants, the main problem is the removal of these compounds. However, other features are present, linked to the health of the population living in its proximity. This Special Issue reports experimental run findings with the aim of removing heavy metals and/or metalloids from soil, making use of challenging techniques, and also demonstrating approaches for the assessment of the risks to human health.

Fulvia Chiampo, Massimo Zacchini
Editors

Editorial

Environmental Restoration of Metal-Contaminated Soils

Fulvia Chiampo [1,*] and Massimo Zacchini [2]

[1] Department of Applied Science and Technology, Politecnico di Torino, Corso Duca degli Abruzzi 24, 10129 Torino, Italy
[2] Research Institute on Terrestrial Ecosystem, National Research Council of Italy, Via Salaria Km 29,300, 00015 Rome, Italy; massimo.zacchini@cnr.it
* Correspondence: fulvia.chiampo@polito.it; Tel.: +39-011-090-4685

Citation: Chiampo, F.; Zacchini, M. Environmental Restoration of Metal-Contaminated Soils. *Appl. Sci.* **2021**, *11*, 10805. https://doi.org/10.3390/app112210805

Received: 14 September 2021
Accepted: 5 November 2021
Published: 16 November 2021

Publisher's Note: MDPI stays neutral with regard to jurisdictional claims in published maps and institutional affiliations.

Copyright: © 2021 by the authors. Licensee MDPI, Basel, Switzerland. This article is an open access article distributed under the terms and conditions of the Creative Commons Attribution (CC BY) license (https://creativecommons.org/licenses/by/4.0/).

The growing industrialization of the last two centuries has improved life to a great extent in the countries where it occurred. However, several drawbacks were derived. Among them, many industrial activities gave and have given origin to dramatic environmental impacts on air, soil, and water.

Concerning soil, the most frequent contamination factor is represented by heavy metals. These pollutants derive from the metal industry, but also agriculture, mining, and waste disposal. Unlike other contaminants, they can persist in the environment for a long time, and this increases the probability to be distributed through the soil layers and transferred to groundwater, with the consistent risk to enter the food chain by water, crops, vegetables, meat, and fish.

Each metallic element has a level of toxicity responsible for health impacts over short, medium, and long times. When the metal concentration is known, the toxicological risk can be assessed by literature correlations, usually based on experimental data. When the risk is considered not acceptable for human health, the request for the reclamation of the polluted sites becomes extremely urgent. The management of a polluted site reclamation is considered one of the most challenging environmental issues to address, as it involves multiple inter-related factors. Several techniques have been developed to remove metals from soil. Traditional soil remediation methods, relying on physical and chemical techniques, have been recently supplemented by those based on biological processes, also called Nature-Based Solutions (NBS).

The multifaceted problems linked to soil heavy metal contamination require an integrated approach involving different expertise and proficiency. In this regard, the current Special Issue aims to focus and highlight the features of this multidisciplinary topic, thanks to the contribution of researchers with different backgrounds to combine knowledge from many disciplines.

A review [1] summarizes these features, showing the relevance of each. This manuscript also describes the main techniques that can be applied for the remediation of metal-contaminated soils. The study highlights the need for tighter cooperation between research and companies involved in remediation, to publish and disseminate the results from experience on a larger scale than the experimental/pilot one.

The first research manuscript presented in the Special Issue checks two methods to consider and assess the risk on human health due to soil ingestion, with a particular focus on bioaccessibility [2]. In this study, Zingaretti and Baciocchi studied the bioaccessible concentration of some metals achieved by two extraction methods, namely, the Unified BARGE Method (UBM) and the Simple Bioaccessibility Extraction Test (SBET). Notwithstanding the lower complexity of the second method, the results were similar, demonstrating that the SBET could be used for screening scopes, while the UBM can be adopted to obtain more accurate data. Then, they used these data to calculate the bioaccessible concentration and the cleanup goal and evidenced the need to include the bioaccessibility into the human health risk assessment (HHRA).

The manuscript by Cameselle et al. [3] presents the results achieved by the electrokinetic removal of heavy metals targeting the remediation of contaminated soils. The experimental runs compared the effect of pH on solubilization and transport of the metallic elements to enhance their removal. EDTA and citric acid were used to change pH. For four metals (Cd, Co, Cu, Zn), the long treatment time (65 days) and high citric acid concentration (0.5 M) resulted in their 70–80% removal, whereas for Cr and Pb, the right operative conditions must be found. As a whole, the results are encouraging for future studies.

Among the techniques tested and successfully applied to restore and clean up soils, sorption constitutes a good and cheap solution. In particular, the use of natural sorbents is emphasized, both for their low cost and the removal efficiency demonstrated in several studies. In the manuscript of Yurak et al. [4], peat, diatomite, vermiculite, and their mixtures were applied to remove one metalloid and five metals (As, Cd, Cr(III), Cr(VI), Cu, Pb). The best result for the removal efficiency was achieved with granular peat–diatomite, followed by large-fraction vermiculite, medium-fraction vermiculite, non-granular peat–diatomite, and diatomite. Unfortunately, one drawback was evidenced: the removal efficiency decreased with time.

Amirahmadi et al. [5] studied the metal adsorption by biochar in mining areas to remediate these heavily polluted soils. They investigated the Cd concentration and bioavailability in pots with loamy soil, monitoring the growth of oak seedlings in the presence of rice husk biochar. The results demonstrated that at the highest tested biochar addition (5% by weight), the bioavailability was always lower than in pots without biochar and also when the Cd concentration reached 50 mg kg^{-1}.

The uptake of one metalloid and some heavy metals (As, Cd, Ce, Co, Cr, Cu, Fe, La, Mn, Ni, Pb, V, Zn) by edible plants in urban soils is the main topic of the manuscript presented by Gaggero et al. [6]. They studied metal uptake in two common vegetables, namely, *Lactuca sativa* and *Brassica oleracea*, grown in contaminated soils, with and without soil amendment deriving from biodegradable wastes. Specifically, the authors analyzed the aerial parts and roots of these vegetables. Then, the results were compared with the ones observed in uncontaminated soil. It was shown that the plants grown in contaminated soils absorbed the toxic elements. When soil amendment was used, the toxic element accumulation was found mainly in the roots, with a limited amount of pollutants reaching the aerial parts (i.e., the edible parts).

As the last manuscript, but of no minor relevance, Han et al. [7] present the study carried out on the flow polluted with copper through a permeable membrane in the filters for contaminant removal, as occurs in Korean landfills. The experimental runs were performed in hydrophobic-coated capillary tubes to simulate the landfill in which similar flow conditions can be present. The results showed that in the center of the tube, the flow rate was always higher than near the surface, where the hydrophobic condition slowed down the hydrophilic contaminant. This also occurred when the pH was changed, namely when it increased from 4 to 10. The experimental data were modeled by computational fluid dynamics (CFD). From the applicative point of view, this means an effect by the pH condition, in terms of slowing down the flow rate by the hydrophobic surface, giving the opportunity of selective remediation.

Altogether, the Special Issue collected data and results coming from different features of the same environmental issue, namely the remediation of soils polluted with heavy metals.

All the studies evidenced the need to further investigate such features. Moreover, all the manuscripts highlighted the crucial point to transfer the current data to a larger scale, to obtain robust data and information for a real-scale application. In other terms, tighter cooperation between research and the industrial world is required.

Author Contributions: The Authors have given the same contribution to all the steps during the preparation of this manuscript. They have read and agreed to its published version. All authors have read and agreed to the published version of the manuscript.

Funding: This research received no external funding.

Acknowledgments: The Guest Editors would like to thank both the authors for their contribution to the Special Issue and the reviewers for their time spent to enhance the quality of the manuscripts and journal. A special thank to Christy Cui and the Editorial Office for their support.

Conflicts of Interest: The authors declare no conflict of interest.

References

1. Raffa, C.; Chiampo, F.; Shanthakumar, S. Remediation of Metal/Metalloid-Polluted Soils: A Short Review. *Appl. Sci.* **2021**, *11*, 4134. [CrossRef]
2. Zingaretti, D.; Baciocchi, R. Different Approaches for Incorporating Bioaccessibility of Inorganics in Human Health Risk Assessment of Contaminated Soils. *Appl. Sci.* **2021**, *11*, 3005. [CrossRef]
3. Cameselle, C.; Gouveia, S.; Cabo, A. Enhanced Electrokinetic Remediation for the Removal of Heavy Metals from Contaminated Soils. *Appl. Sci.* **2021**, *11*, 1799. [CrossRef]
4. Yurak, V.; Apakashev, R.; Dushin, A.; Usmanov, A.; Lebzin, M.; Malyshev, A. Testing of Natural Sorbents for the Assessment of Heavy Metal Ions' Adsorption. *Appl. Sci.* **2021**, *11*, 3723. [CrossRef]
5. Amirahmadi, E.; Hojjati, S.M.; Kammann, C.; Ghorbani, M.; Biparva, P. The Potential Effectiveness of Biochar Application to Reduce Soil Cd Bioavailability and Encourage Oak Seedling Growth. *Appl. Sci.* **2020**, *10*, 3410. [CrossRef]
6. Gaggero, E.; Calza, P.; Fabbri, D.; Fusconi, A.; Mucciarelli, M.; Bordiglia, G.; Abollino, O.; Malandrino, M. Assessment and Mitigation of Heavy Metals Uptake by Edible Vegetables Grown in a Turin Contaminated Soil Used as Vegetable Garden. *Appl. Sci.* **2020**, *10*, 4483. [CrossRef]
7. Han, J.-G.; Jung, D.; Lee, J.-Y.; Kim, D.; Hong, G. A Study on the Flow Characteristics of Copper Heavy Metal Microfluidics with Hydrophobic Coating and pH Change. *Appl. Sci.* **2021**, *11*, 4328. [CrossRef]

Review

Remediation of Metal/Metalloid-Polluted Soils: A Short Review

Carla Maria Raffa [1], Fulvia Chiampo [1,*] and Subramanian Shanthakumar [2]

[1] Department of Applied Science and Technology, Politecnico di Torino, Corso Duca degli Abruzzi 24, 10129 Torino, Italy; carla.raffa@polito.it
[2] Department of Environmental and Water Resources Engineering, School of Civil Engineering, Vellore Institute of Technology, Vellore 632014, India; shanthakumar.s@vit.ac.in
* Correspondence: fulvia.chiampo@polito.it; Tel.: +39-011-090-4685

Citation: Raffa, C.M.; Chiampo, F.; Shanthakumar, S. Remediation of Metal/Metalloid-Polluted Soils: A Short Review. *Appl. Sci.* 2021, 11, 4134. https://doi.org/10.3390/app11094134

Academic Editor: Eduardo Ferreira da Silva

Received: 4 March 2021
Accepted: 27 April 2021
Published: 30 April 2021

Publisher's Note: MDPI stays neutral with regard to jurisdictional claims in published maps and institutional affiliations.

Copyright: © 2021 by the authors. Licensee MDPI, Basel, Switzerland. This article is an open access article distributed under the terms and conditions of the Creative Commons Attribution (CC BY) license (https://creativecommons.org/licenses/by/4.0/).

Abstract: The contamination of soil by heavy metals and metalloids is a worldwide problem due to the accumulation of these compounds in the environment, endangering human health, plants, and animals. Heavy metals and metalloids are normally present in nature, but the rise of industrialization has led to concentrations higher than the admissible ones. They are non-biodegradable and toxic, even at very low concentrations. Residues accumulate in living beings and become dangerous every time they are assimilated and stored faster than they are metabolized. Thus, the potentially harmful effects are due to persistence in the environment, bioaccumulation in the organisms, and toxicity. The severity of the effect depends on the type of heavy metal or metalloid. Indeed, some heavy metals (e.g., Mn, Fe, Co, Ni) at very low concentrations are essential for living organisms, while others (e.g., Cd, Pb, and Hg) are nonessential and are toxic even in trace amounts. It is important to monitor the concentration of heavy metals and metalloids in the environment and adopt methods to remove them. For this purpose, various techniques have been developed over the years: physical remediation (e.g., washing, thermal desorption, solidification), chemical remediation (e.g., adsorption, catalysis, precipitation/solubilization, electrokinetic methods), biological remediation (e.g., biodegradation, phytoremediation, bioventing), and combined remediation (e.g., electrokinetic–microbial remediation; washing–microbial degradation). Some of these are well known and used on a large scale, while others are still at the research level. The main evaluation factors for the choice are contaminated site geology, contamination characteristics, cost, feasibility, and sustainability of the applied process, as well as the technology readiness level. This review aims to give a picture of the main techniques of heavy metal removal, also giving elements to assess their potential hazardousness due to their concentrations.

Keywords: heavy metals and metalloids; removal techniques; contaminated soil

1. Introduction

Soil contamination by heavy metals and metalloids is a problem that all countries in the world are facing. In recent years, the attention to heavy metal pollution has increased since they are non-biodegradable; they accumulate in soil causing damage to the environment, animals, and humans for a long time [1]. Multiple health effects are associated with exposure to heavy metals and metalloids: kidney and bone problems, neurobehavioral and developmental disorders, blood pressure problems, and tumor formation.

The problem becomes relevant when the concentration of heavy metals in the soil is high. Around the world, it is estimated that the number of sites with soil contaminated by heavy metals and metalloids is around five million [2], and anthropological activities are usually the origin of this pollution. Most of these heavy-metal-contaminated sites are in developed countries, such as the United States of America, Australia, the European Union State Members, and China. For example, in the USA, around 6000 km^2 have been contaminated by heavy metals/metalloids; 250,000 sites are polluted in the European Union; and

810,000 km² of farmland in China have been contaminated by heavy metals/metalloids [3]. The study of Cheng et al. [4] reports that more than 30,000 tons of chromium and 800,000 tons of lead have been released into the environment globally in the past half-century. Different studies have focused on determining the concentrations of heavy metals and metalloids in the European Union [5], China [6–8], and Brazil [9,10]. The amounts of these pollutants in soils are different and worldwide average concentrations depend on the type of soil, environmental conditions, and the distance from the contamination source.

In this review, the aim is to give information about the elements (metals and metalloids) that contribute to environmental pollution, due to their toxicity, persistence in the environment, and bioaccumulation in nature, and depict the main techniques that have demonstrated efficacy to remove these elements so far.

This review is focused on the removal of heavy metals, analyzing: (1) the toxicity effects of these pollutants on living organisms; (2) the legislation in force in some developed countries; and (3) the main remediation processes for contaminated soils.

Over the years, many authors have studied this topic, due to its great importance at the global level. Among the existing reviews, one was published by Dhaliwal et al. in 2020 [11]. However, it was mainly focused on biological remediation techniques at the laboratory level, and did not give detailed information on the other issues, which are important matters. To overcome this limitation, these issues have been developed in the current review. In addition to this, the last section of this review provides some case studies, and this constitutes a novelty compared to previous reviews, which have usually been limited to experimental studies and data, with limited (or without) links to their application.

These techniques are based on physical, chemical, and biological processes [12], and they can be classified as:

- physical methods (landfilling and leaching, excavation, soil washing, calcination) that permit high removal efficiency and the treatment of large quantities of soil, but are expensive;
- chemical methods (soil washing, electrochemical remediation, adsorption) that are very effective, but can be a source of new chemical contaminants introduced into soils, for example, in soil washing;
- physical–chemical processes (ion-exchange, precipitation, reverse osmosis, evaporation, and chemical reduction) which are simple and easy to apply, but have a high-cost burden;
- bioremediation processes (bioventing, biosparging, bioaugmentation, biostimulation) which are environmentally friendly and cost-effective, but the degradation time is slower than in other treatments.

2. Origin, Characteristics, and Properties of Heavy Metals and Metalloids

The term "heavy metals" is used to refer to metals and metalloids present in the environment and having a density higher than 5000 kg·m^{-3} and an atomic mass higher than 20. Based on this definition, 51 elements of the periodic table are considered heavy metals/metalloids [13].

Their mobility and bioavailability in soil are due to their chemical characteristics and those of soil. The pH, surface properties of the adsorbents, presence of cations, and anions affect the interactions between soil components and metals/metalloids [14].

The most common heavy metals and metalloids in the environment are chromium, manganese, nickel, copper, zinc, cadmium, lead, and arsenic. Chromium, copper, zinc, cadmium, lead, mercury, and arsenic are the most toxic [15].

Usually, they found in the weathering of underlying bedrock, and they are present as ores (sulfides of Pb, Co, Fe, As, Pb, Zn, Ag and Ni, and oxides of Se, Al, Mn, and Sb). In soils, normally, sulfides of arsenic, mercury, lead, cadmium naturally occur together with sulfides of copper (chalcopyrite, $CuFeS_2$) and iron (pyrite, FeS_2) [3].

The environmental problem is mainly due to the anthropological activities that cause an increase in heavy metal and metalloid concentrations, especially in the refining and mining of ores, pesticide applications, fertilizer industries, and solid wastes [16]. The anthropogenic sources of metals and metalloids can be dived into five groups, as shown in Figure 1.

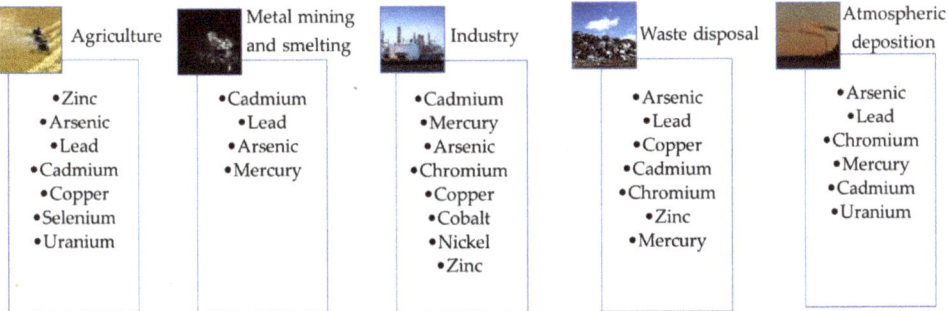

Figure 1. Anthropogenic sources of polluting metals and metalloids.

Heavy metals and metalloids are used in different sectors, increasing market demand and world production. Copper, selenium, zinc, iron, vanadium, and manganese, in trace amounts, are essential for various biological processes, such as in respiration systems, biosynthesis of complex compounds, nervous system, regulation, and functioning of enzymes. Iron, zinc, tin, lead, copper, and tungsten have an important role in electronic devices, especially in the realization of semiconductors [17]. In Table 1, the characteristics of the most toxic heavy metals and metalloids are summarized [18,19].

Table 1. Heavy metals and metalloids characteristics.

Element	Chemical and Physical Properties	Application	World Production (ton·y^{-1})
Chromium (Cr)	Density: 7190 kg·m^{-3} Atomic mass: 51.99 Heat of fusion: 21.00 kJ·mol^{-1}	Industrial application, alloys, tanning agents, paint pigments, catalysts, photography.	15,000,000 (year: 2017)
Copper (Cu)	Density: 8960 kg·m^{-3} Atomic mass: 63.55 Heat of fusion: 13.26 kJ·mol^{-1}	Electrical and electronics, transport equipment, construction, industrial machinery, pesticides.	20,000,000 (year: 2017)
Zinc (Zn)	Density: 7140 kg·m^{-3} Atomic mass: 65.38 Heat of fusion: 7.32 kJ·mol^{-1}	Paints, rubber, cosmetics, pharmaceuticals, plastics, inks, soaps, batteries, textiles, and electrical equipment	13,500,000 (year: 2019)
Cadmium (Cd)	Density: 8650 kg·m^{-3} Atomic mass: 112.41 Heat of fusion: 6.21 kJ·mol^{-1}	Electroplating, paint pigments, plastics, silver–cadmium batteries, coating operations, machinery and baking enamels, photography, television phosphors.	24,670 (year: 2019)
Lead (Pb)	Density: 11,340 kg·m^{-3} Atomic mass: 207.2 Heat of fusion: 4.77 kJ·mol^{-1}	Electrical accumulators and batteries, building construction, cable coatings ammunition.	11,600,000 (year: 2018)
Mercury (Hg)	Density: 13,530 kg·m^{-3} Atomic mass: 200.59 Heat of fusion: 2.29 kJ·mol^{-1}	Dental preparations, thermometers, fluorescent and ultraviolet lamps, pharmaceuticals, fungicides, industrial process waters, seed dressings.	4000 (year: 2019)
Arsenic (As)	Density: 5730 kg·m^{-3} Atomic mass: 74.92 Heat of fusion: 24.44 kJ·mol^{-1}	Pesticides, pharmaceuticals, alloys, semiconductors.	33,000 (year: 2019)

It is evident that these elements are used in all fields and that annual production around the world is very high, especially chromium, copper, zinc, and lead. The major producing countries are China, Peru, Australia, the United States of America, Russia, and Mexico.

3. Toxicology

The heavy metals and metalloids accumulate in the environment where anthropogenic sources increase the background quantity, and this effect becomes risky when the natural concentrations are such that they cause damage to living organisms. Their toxicity is relevant at low concentrations, as demonstrated by the limit concentrations imposed by legislative acts (Table 2).

Table 2. Limit concentrations of heavy metals and metalloids in the soil.

Country/Organization	Type of Soil	Hg	Cd	Pb	Cr(VI)	Ni	Unit	Ref.
WHO	Agricultural soil	0.08	0.003	0.1	0.1	0.05	ppm	[20]
China	Agricultural soil	0.3–1.0	0.3–0.6	80	150–300	40–60	ppm	[20]
US	Agricultural soil	1.0	0.43	200	11	72	ppm	[21]
Italy	Residential soil	1	2	100	2	120	mg·kg^{-1}	[22]
	Industrial soil	5	15	1000	15	500	mg·kg^{-1}	
Finland	Threshold value	0.5	1	60	100	50	mg·kg^{-1}	[23]
	Lower guideline value	2	10	200	200	100	mg·kg^{-1}	
	Higher guideline value	5	20	750	300	150	mg·kg^{-1}	
Canada	Agricultural soil	0.8	3	200	250	100	mg·kg^{-1}	[24]
Germany	Agricultural soil	5	5	1000	500	200	mg·kg^{-1}	[2]
Spain	Soil pH < 7	1	1	50	100	30	mg·kg^{-1}	[25]
	Soil pH > 7	1.5	3	300	150	112	mg·kg^{-1}	

The concentration limits vary from country to country as there is no international regulation. For example, in the European Union, this topic is not ruled by a directive, enforcing different threshold limits in the 27 countries. Furthermore, the limit value depends on the element type and the land use of the soil. The mobility of these elements in soil makes them available to plants and consequently, they enter the food chain [26]. For this reason, it is important to evaluate the diffusion of toxic elements in soil.

3.1. Presence and Distribution into the Environment

Several studies report on the distribution of heavy metals and metalloids in soil [9,10,27,28]. Some soil properties facilitate the accumulation of these pollutants and increase their concentrations. pH controls the adsorption and solubilization; if the soil pH is acidic, the solubility and the bioavailability of heavy metals/metalloids increase [29]. Organic matter content (OMC) is another factor that influences their presence since organic substances with a high molecular mass have a high affinity with heavy metals and form water-insoluble metal complexes [30]. In general, the distribution of heavy metals and metalloids is high in topsoil and decreases as the depth increases [31].

Some studies evaluated their concentrations in soils close to industrial activities, and some examples are reported in Table 3.

Table 3. Concentration of heavy metals/metalloids in polluted sites.

Country	Proximity to Activity Sources	Soil	Element	Concentration (mg·kg^{-1})	Ref.
China (Feng County)	Pb/Zn smelter	A = County Seat (pH 8.5) B = River basin (pH 8.0–8.5) C = Smelter area (pH 8.6)	Cd	A: 6.7 B: 0.8–2.7 C: 57.6	[31]
			Cu	A: 25.0 B: 21.9–30.2 C: 36.9	
			Ni	A: 46.0 B: 39.4–46.3 C: 49.3	
			Pb	A: 50.0 B: 30.0–70.0 C: 148.0	
			Zn	A: 900.0 B: 300.0–400.0 C: 2079	
China (Shuozhou)	Pingshuo open pit mine		Cd	0.117	[8]
			Hg	0.03	
			As	9.629	
			Pb	21.328	
			Cr	55.609	
China	Pb smelter	Distance from smelter D = 1 km E = 3 km F = 6 km	Cd	D: 4.5 E: 2.0 F: 1.0	[6]
			Cu	D: 45.0 E: 35.0 F: 32.0	
			Pb	D: 350.0 E: 180.0 F: 175.0	
			Zn	D: 128.0 E: 82.0 F: 80.0	
China (Changshu City, Jiangsu Province)	Primary, secondary, and tertiary industries	Gleyic clayey paddy soil	Mg	0.22	[32]
			K	1.64	
			V	82.77	
			Se	0.12	
			Mn	347.77	
			Fe	1.16	
			Co	12.76	
			Sb	4.14	
			Pb	31.41	
			Cu	31.60	
			Zn	61.13	
			As	7.46	
			Cr	86.38	
			Cd	0.11	
			Ni	34.93	

Table 3. *Cont.*

Country	Proximity to Activity Sources	Soil	Element	Concentration (mg·kg^{-1})	Ref.
Pakistan (Swabi)		G = Depth (cm) 0–15 Organic matter (%) 0.35–2.30 pH 7.21–9.21 EC (dS·m^{-1}) 0.13–0.56 CaCO$_3$ (%) 5.89–16.65 H = Depth (cm) 15–30 OMC (%) 0.21–1.52 pH 7.32–8.88 EC (dS·m^{-1}) 0.18–0.86 CaCO$_3$ (%) 6.56–17.81	Cu	G: 2.33–19.15 H: 1.32–14.11	[33]
			Fe	G: 8.23–36.89 H: 8.22–30.95	
			Zn	G: 8.26–26.55 H: 7.77–24.20	
			Cd	G: 0.01–0.16 H: 0.01–0.08	
			Co	G: 0.8–6.99 H: 0.33–5.46	
			Ni	G: 0.46–22.21 H: 0.42–21.9	
			Cr	G: 0.23–8.02 H: 0.54–5.11	
			Pb	G: 0.4–2.23 H: 0.08–1.99	
Colombia (Sinú River Basin)		Agricultural soil	Cu	1149	[29]
			Ni	661	
			Pb	0.071	
			Cd	0.040	
			Hg	0.159	
			Zn	1365	
Greece (Argolida)		Agricultural soil	Cu	28.64	[34]
			Pb	13.96	
			Zn	45.26	
			Ni	253.7	
			Co	25.05	
			Mn	665	
			As	5.89	
			Cd	0.26	
			Cr	138.4	
			Fe	2.90	
			P	0.039	
			K	0.239	

Very often, in industrial areas, the concentrations of heavy metals and metalloids are higher than the threshold limits allowed by national legislation. Some authors compared the concentration values of the same soil, before and after the presence of anthropological sources. They demonstrated that an increase of concentration was precisely due to these activities [8]. In agricultural soils, high concentrations are due to the use of fertilizers and pesticides [29].

Regarding the European Union, an interesting study was done by Tòth et al. [5]; where they monitored As, Cd, Cr, Cu, Hg, Pb, Zn, Sb, Co, and Ni concentrations. Results showed that, in most soil samples taken from various regions, concentrations were above the threshold limits, both on their entire land area and on agricultural lands. In Western Europe and Mediterranean regions, the concentrations were higher than those in North-Eastern Europe and Eastern-Central Europe. The measured values of arsenic exceeded threshold values by 10% to 90%, and the most polluted regions were France, Italy, and Spain. For cadmium, the values were 10% to 70% above threshold values, and the highest concentrations were found in Ireland and Greece. A few European regions were found with cobalt values higher than the threshold values, among them, France and Greece. Zinc, mercury, and copper values were not higher than 10% of guideline values and most regions had concentrations below 2% of the threshold limits. Nickel and antimony contaminations were significant, especially in Italy and Greece since the concentrations exceeded 90% of the admissible values.

3.2. Health Impacts Over Short, Medium, and Long Term

The high amounts of these elements in soil lead to the deterioration of agricultural land, eutrophication, and the absorption of toxic substances. Direct effects of contamination are a reduction in the quality of agricultural soils, including phytotoxicity at high concentrations, the preservation of soil microbial processes, and the transfer of zootoxic elements to the human diet [35].

The health of living organisms is damaged, especially in the long term. The principal effects on health can be reduced growth and development, cancer, organ damage, nervous system damage, and, in extreme cases, death [36]. Toxicity depends on the type and form of the element, dose, route of exposure, age, genetics, and nutritional status of the exposed individuals [37].

The United States Environmental Protection Agency (EPA) establishes limits for some heavy metals and metalloids; surpassing these values has toxic effects on human health [38].

It is possible to summarize the principal effects of the most toxic metals/metalloids when the admissible concentration is exceeded:

- Chromium (Cr): it is found in the environment in different oxidation states (-2 to $+6$), but the most stable forms are trivalent (III) and hexavalent (VI). Chromium(VI) is more absorbed than chromium(III) by the human body through inhalation, ingestion, and dermal contact, due to its high solubility and mobility. Chromium(VI) enters into the cell via a nonspecific anion channel via facilitated diffusion, while chromium(III) enters by passive diffusion or phagocytosis. The main organs that are influenced are the liver, kidney, spleen, and bone. The hexavalent form can easily penetrate red blood cells [39]. The main toxic effects of chromium for humans are ulcers, dermatitis, perforation of the nasal septum, and respiratory cancer. In soil, chromium alters the structure of microbial communities and reduces their growth [40].
- Copper (Cu): this element is essential in many biological processes (oxidation, photosynthesis, and carbohydrate, protein, and cell wall metabolism) [41]. Excessive concentration of copper leads to the formation of free radical species that damage the cell and inactivate some enzymes, threatening the environment, microbes, and human health [42]. In particular, the accidental ingestion of copper may cause nausea, vomiting, and abdominal pain, whereas a prolonged exposure leads to chronic effects, involving the liver and kidneys [36]. In the plants, copper accumulates in the roots, reducing their growth and the ability to absorb other trace elements useful for plant development [43].
- Zinc (Zn): is a trace element essential for all organisms, important in nucleic acid and protein metabolism, in cell growth, division, and function [44]. Excessive concentration in food or potable water may cause vomiting, muscle cramps, and renal damage [45]. In plants, a high concentration of zinc leads to a decrease in growth (both roots and shoots) and development of the plant, chlorosis, alteration in metabolism processes, and induction of oxidative damage [46].
- Cadmium (Cd): has eight stable isotopes, the most common are ^{112}Cd and ^{114}Cd. Cadmium forms various complexes with amines, sulfur, chlorine, and chelates. Depending on the form, it has different clinical manifestations and toxic effects. Cadmium interferes with cell proliferation, differentiation, apoptosis, and DNA repair mechanism. The common clinical effects are skeletal demineralization, kidney, and liver problems [47]. Excess accumulation in plants can influence both photosynthesis and respiration, transport, and assimilation of mineral nutrients, affecting plant growth and development [48].
- Lead (Pb): lead poisoning occurs mainly by the ingestion of food and water. It is quickly absorbed into the bloodstream, damaging various systems [49]. Different studies have reported the dangerous effects of lead on the neurologic system, such

as irritability, agitation, headaches, confusion, ataxia, drowsiness, convulsions, and coma, and has an effect on renal functions, the body development, and the lymphatic system [50].

- Mercury (Hg): in the environment, mercury can be present in the form of both organic and inorganic (Hg, Hg_2^{2+}, Hg^{2+}) compounds. Mercury tends to deposit in many parts of the human body, damaging the brain, thyroid, breast, myocardium, muscles, liver, kidneys, skin, and pancreas. Among them, the nervous system is the most affected [51]. Inorganic mercury can inhibit the activity of enzymes in the body and destroy the normal metabolism of cells. The organic form has a great negative effect on brain function and can enter through the food chain [52].
- Arsenic (As): it is present in the environment in organic and inorganic forms, the most common and toxic forms are arsenate and arsenite. Arsenate can cause damage to the plant since it affects phosphate metabolism, while arsenite binds to sulfhydryl groups of proteins, interfering with their structures and functions [53,54]. Chronic exposure to this metalloid causes cutaneous lesions, such as melanosis (hyperpigmentation), keratosis, and leukomelanosis (hypopigmentation); lung, bladder, liver, and kidney cancers; ischemic heart diseases, impaired cognitive abilities, motor functions, and hormonal regulations [55].

If the heavy metal and metalloid concentrations in soil are known, some methods to assess the associated risks are available in the scientific literature. For example, the Hakanson method permits to calculate the potential ecological risk index (RI) [56–58]:

$$RI = T_r \cdot C_f = T_r \cdot \frac{C}{C_n} \quad (1)$$

where:

T_r is the toxic response factor of the element, which reflects the toxicity level and sensitivity of organisms to it [59]. This factor is calculated with an empirical formula, considering the effects that the element can cause in microorganisms. For Cr, Ni, Cu, As, Cd, and Pb, T_r is equal to 2, 6, 5, 10, 30, and 5, respectively;

C_f is the pollution index of the element, the ratio between the measured concentration of an element in the tested area, C, and the background (to say, natural) concentration of an element in the tested area, C_n.

The RI value reflects the state of the soil under examination, according to these assessments:

- $RI < 150$: the ecological risk is low;
- $150 \leq RI \leq 300$: the ecological risk is moderate;
- $300 \leq RI \leq 600$: the ecological risk is considerable;
- $RI \geq 600$: the ecological risk is very high.

Another method that defines the hazard quotient (HQ) [58] is expressed as:

$$HQ = \frac{C}{C_b} \quad (2)$$

where:

C is the measured concentration of an element in a study area;

C_b is the benchmark concentration of the element (the limit value concentration given by current legislation).

If the hazard quotient for the test is lower than 1 ($HQ < 1$), negative effects and ecological risks are highly unlikely. Vice versa, if $HQ > 1$, the soil must be remediated.

4. Removal Techniques

The removal techniques developed over the years can be divided into three categories: physical, chemical, and biological processes [60,61]. Their aim is the complete removal of contaminants or the transformation of contaminants into less dangerous forms [62]. They differ by in terms of the mechanism applied to remove/degrade the pollutants from soil.

The techniques can be also characterized by the location where they are applied, namely:

- *in situ*: the treatment is carried out directly on the site where the pollution is present, and there is no need to move/excavate the soil;
- *on site*: the soil is removed and processed on site surrounding the polluted area. The technique can be carefully monitored and kept under control;
- *ex situ*: the remediation occurs at a site far from the polluted area, and this entails soil excavation, its transport to a processing plant, and often transport back to the original site.

For some techniques, the operation can be carried out at all kinds of locations, whereas for others it can occur only *in situ*. Each location involves different features and impacts that are typical of the considered polluted site. Therefore, a detailed discussion with an assessment of pros and cons is impossible.

Table 4 reports a summary of the techniques and their main characteristics.

Table 4. Removal techniques for heavy metals/metalloids.

Technique	Studies	Maximum Removal	Advantages	Disadvantages	Location
Physical processes Removal through physical operations	Physical separation [63–66] Thermal treatment [67,68] Vitrification [69]	95%	High efficiency Simplicity Rapidity	Cost	A, B, C A, B, C A
Chemical processes Removal through chemical operations	Stabilization [70,71] Treatment with nanoparticles [72–81] Stabilization/solidification [82] Chemical soil washing [83–85] Electrochemical remediation [86–90]	90%	High efficiency Simplicity Rapidity	Cost Changes in the physicochemical soil properties	A, B, C B, C A B, C A
Biological processes Removal of pollutants by the microbial activity of microorganisms, or plants, or their combination	Biosorption [91–99] Bioleaching [100,101] Phytoremediation [102–117]	96%	Cheapness Ecofriendly Simplicity	Slow High efficiency only with low pollutant concentration	A, B, C B, C A

A: *in situ* treatment; B: *on site* treatment; C: *ex situ* treatment.

4.1. Physical Processes

Physical technologies permit removal through physical mechanisms such as physical separation, soil replacement, thermal treatment, or vitrification (Figure 2). These methods allow us to have a high removal efficiency, but they are rather expensive.

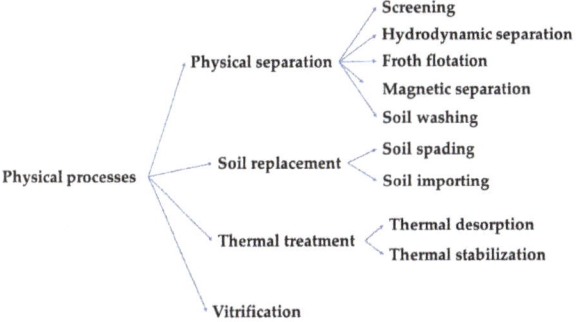

Figure 2. Physical processes to remove heavy metals and metalloids.

4.1.1. Physical Separation

Physical separation permits the removal of heavy metal/metalloid particles from contaminated soils at a large scale [64]. This technique exploits the differences in certain physical characteristics between metal particles and soil, such as size, density, magnetism, and hydrophobic surface properties [118]. Physical separation does not alter the chemical properties of soil, such as pH, elemental composition, and OMC [63].

Depending on the particle properties, physical separation can be done by [118]:

- Screening: the particle separation occurs on the basis of particle size, by passing (or not) through screen holes.
- Hydrodynamic separation: this operation exploits the different settling velocities of particles in a water flow; as an alternative, the different effects of centrifugal force can be applied.
- Froth flotation: the method exploits the differences in hydrophobic properties to separate metal-bearing particles from the soil matrix through air bubbles injected in a soil slurry.
- Magnetic separation: separates particles based on their different magnetic properties.
- Soil washing: the contaminates are desorbed and extracted from soil with an extractant solution.

Boente et al. [63] applied physical techniques to separate arsenic, copper, mercury, lead, and antimony from a brownfield affected by pyrite ash disposal. The used methods were high-intensity magnetic separation and hydrodynamic centrifugal separation. Magnetic separation was tested both under wet conditions, this is to say that the soil was mixed with water to form a slurry, and under dry conditions, with dried soil. The tests were carried out by changing the soil granulometry. They found that dry magnetic separation was more appropriate for large particle sizes (>250 µm), while the wet process was suitable for all particle sizes, except for clay soils (particle size lower than 63 µm). Hydrodynamic centrifugal separation was efficient for a wide fraction of sand (63–500 µm).

Liao et al. [64] found that the concentrations of heavy metals such as Pb, Cd, and Zn in soil increase with decreasing particle size and when soil has a high content of organic matter or clays. The use of mechanical mixing in soil washing promotes physical contact between contaminated soil particles and the washing liquid. To increase the efficiency of heavy metals removal, ultrasonic mechanical soil washing was also studied [65].

The advantages of physical methods are their high efficiency, simplicity, and rapidity of application, but they have a high cost, and they can change some soil properties; for example, soil texture or particle size, causing a deterioration in soil fertility [66].

4.1.2. Soil Replacement

This technique is indicated for treatment in small-scale applications. The method involves the total or partial replacement of contaminated soil with an uncontaminated one [119]. The replaced soil is treated as waste. It is also possible to decrease heavy metals/metalloid concentration with soil spading or new soil imports. With soil spading, contaminated soil is dug deeply, spaded, and then substituted with clean soil. With importing new soil, clean soil is added and mixed with polluted soil to reduce the heavy metal and metalloid concentrations [120]. Before replacement, the polluted soil area must be isolated from its surroundings through physical barriers to prevent the contamination of the neighboring areas and groundwater. Soil replacement is onerous and expensive but suitable for small areas that are severely contaminated.

4.1.3. Thermal Treatment

Thermal desorption, such as calcination, is the process of heating a medium under controlled temperature to eliminate a volatile substance. The main techniques for heating soil are conductive heating, electrical resistive heating, steam-based heating, and radio-frequency heating [60]. Volatile pollutants are desorbed from soil and are collected using vacuum-negative pressure or a carrier gas. The process can occur at high-temperature desorption (320–560 °C) or low-temperature desorption (90–320 °C), depending on the pollutant boiling point [11]. For example, this technique is widely used to treat contaminated soils with mercury. This metal boils at 357 °C, therefore thermal desorption is a good option to remove this metal from soil [68].

Wang et al. [67] studied thermal stabilization to reduce the mobility and availability of zinc and copper in contaminated soils. The metals are fixed in the treated soil. They found that the residual concentrations in tested soils decreased as the temperature increase to 700 °C.

The advantages of using this method are safety, little secondary pollution production, and less energy consumption compared with other processes [68]. However, it requires high capital costs and gas emission control, but it is only really effective high pollutant concentrations and it can damage soil structure.

4.1.4. Vitrification

Vitrification is a thermal technique at high temperatures (>1500 °C) to reduce the mobility of heavy metals and metalloids by fixing them into vitreous material. During the process, contaminated soil is heated with high-voltage electricity, an external heat source, or via electrical discharge-induced gas plasma. The soil is melted into molten lava, which, after cooling, is transformed into a vitrified structure. The heavy metals/metalloids are encapsulated in this glassy matrix, while the other contaminants are destroyed [69]. Vitrification is an efficient technique, but it is destructive for soil, expensive, and very complex. Moreover, it is not applicable to soil with high organic matter content, high moisture content, and soil contaminated by volatile or flammable organics.

4.2. Chemical Processes

Chemical techniques exploit chemical phenomena such as ion exchange and chemical reactions to stabilize and fix heavy metals and metalloids into less toxic forms. For these processes, chemical reagents are required. The main chemical processes are shown in Figure 3.

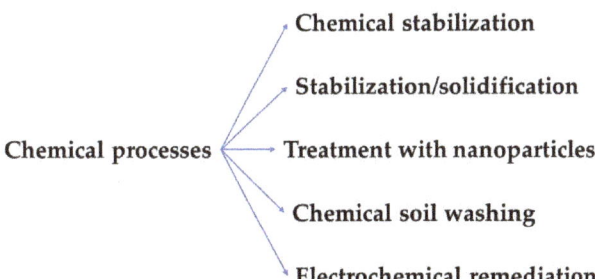

Figure 3. Chemical processes to remove heavy metals and metalloids.

4.2.1. Chemical Stabilization

Chemical stabilization reduces the mobility, bioavailability, and bioaccessibility of heavy metals and metalloids in soil. It consists of adding specific immobilizing agents. These agents promote precipitation, complexation, or adsorption to immobilize the pollutants.

In general, chelating agents with functional groups able to donate nitrogen, oxygen, sulfur, or phosphorus atoms can more likely react with heavy metal ions [82].

As a low-cost alternative, phosphates and carbonates have been demonstrated to be efficient [121].

Several materials including organic ones containing sulfur (hair or cysteine), manganese compounds, zeolite, or iron oxide [70] can also be applied. Nowadays, one of the aims is to use waste-derived resources, that are cost-effective and environmentally-friendly, such as biochar [122,123]. Biochar is a carbonaceous material produced by pyrolysis, that is able to stabilize heavy metals and metalloids [124]. Islam et al. [71] used waste (eggshells and cockle shells) to immobilize Cd, Pb, and Zn in mine tailing soil.

These stabilizers can be sprayed in aqueous solutions over a site if they are water-soluble; alternatively, the chemical agents can be mixed with contaminated soil.

This process aims to immobilize heavy metals [82], but the used chemicals will remain in the soil, and their concentrations will have to be monitored after the remediation.

4.2.2. Nanoparticles

In the last few years, nanotechnologies have been widely applied in many fields, including soil remediation. The synthesis of nanoparticles (D < 100 nm) for heavy metal removal has demonstrated feasibility for cleaning soils. In general, nanoparticles can remove pollutants through different mechanisms: adsorption [125], redox reactions [126], precipitation [127], and co-precipitation [128], all enhanced by a large specific surface area.

Studies making use of different nanomaterials are available. The results evidenced that an excellent removal can be achieved with zero-valent iron nanoparticles (nZVI), where the iron support can have different origins: for example, bentonite [72,73], zeolite [74,75], biochar [76,77], vinegar residues [78], rhamnolipid [79], carboxymethylcellulose [80], and starch [81]. In general, this kind of nanoparticle has shown very good results for the removal of all heavy metals and metalloids.

Recently, attention has been paid to the use of nanoparticles based on a metal–organic framework. This kind of nanoparticle has found wide use in wastewater treatment [129]. However, preliminary tests have also demonstrated positive results for its application to polluted soil remediation [130]. Further studies are needed.

However, nanoparticle techniques are the most controversial, since, at present, the effects on soil properties are not well known, especially in a real long-term scale. Considering the high reactivity of nanoparticles, the main problems to be solved are the potential toxicity of the nanomaterials, the interaction of soil–nanoparticles, the impacts on biodiversity and different ecosystems present in the soil, and the actions to regenerate nanoparticles. This requires a deeper knowledge of the question, given the transfer of this technique to a larger scale.

4.2.3. Stabilization/Solidification

Solidification is a chemical technique where a binding agent, commonly cement, asphalt, fly ash, or clay, is added to a contaminated zone to form a solid block that prevents heavy metals/metalloids from leaching.

The two chemical techniques (stabilization and solidification) can be used together to achieve and maintain the desired physical properties of the soil and chemically stabilize the contaminants in a solid phase. Stabilization/solidification is less harmful to the environment and biota since the chemical reagents remain only in the treated area. The first step is stabilization to reduce the mobility of contaminants with the chemical agents, and then the solidifying agents are added to avoid pollutant diffusion to the surroundings in the future. For the stabilization of heavy metals, traditional additives include cement and silica fume. Liu et al. [82] used a resin (water-soluble thiourea-formaldehyde (WTF)) to adsorb and fix cadmium and chromium. The results showed that this resin is suitable for stabilization, is easier to use than other agents, and is not toxic to indigenous microorganisms.

4.2.4. Chemical Soil Washing

Chemical soil washing permits the elimination of heavy metals/metalloids from contaminated soils by extraction of the pollutants followed by a reaction to produce insoluble compounds. Leaching needs an extractant solution, which can be water, chelators, inorganic and organic acids, or surfactants, also containing the reactant that will permit the production of metal hydroxides, sulfides, carbonates, and phosphates. The derived solid particles can be separated via sedimentation or filtration at the end of the process. The description of the chemical soil washing method is reported, for example, by Zhang et al. [131], who underlined how this technique permits to reduce the most mobile metals. The drawback is the chance to destabilize some strongly bound fractions, compromising the effectiveness of the process itself.

Several agents have been studied. The most-used chelator is ethylenediaminetetraacetic acid (EDTA), which is excellent for the process since it has a strong extraction capacity, but it may cause secondary pollution [83]. EDTA is poorly biodegradable and persistent in soil, and is damaging to soil functions. For this reason, more environmentally friendly agents have been tested.

Feng et al. [83] used ethylenediamine tetra (methylene phosphonic acid) (EDTMP) and polyacrylic acid (PAA) for soil washing, and tried to optimize operative conditions and reduce ecological risks and toxicity. The same was done by Wang et al. [84] who studied four types of acids that are less toxic and more biodegradable than EDTA, namely iminodisuccinic acid (ISA), glutamate–N,N–diacetic acid (GLDA), glucomonocarbonic acid (GCA) and polyaspartic acid (PASP). They demonstrated that these acids are more biodegradable but less effective than EDTA. Zhai et al. [85] coupled soil washing with soil stabilization to improve results. They tested the addition of some reactants (lime, biochar, and carbon black) to reduce the bioavailability of residual metals, and these agents caused less damage to soil microbial communities.

Chemical soil washing is an efficient, rather fast, and widely used method. The main drawbacks are damage to soil properties and the creation of secondary pollution due to the presence of chemical agents.

4.2.5. Electrochemical Remediation

Electrochemical remediation promotes the migration of heavy metals and metalloids to oppositely charged electrodes under a direct-current electric field. Electrokinetic technology involves the transport of charged chemical species in fluid (electromigration), the motion of a fluid (electro-osmosis), the movement of charged particles (electrophoresis), and the chemical reactions associated with an electric current (electrolysis) [132]. Electrochemical remediation technologies have already been described by several authors, such as Reddy and Cameselle [133]. Sun et al. [134] estimated a model to simulate and describe heavy metal transport in soil under an electrical field.

Electrodes are generally immersed into wells containing an electrolytic solution and they are inserted into the area to be treated (Figure 4). An electric field gradient is generated so that metal ions migrate towards the oppositely charged electrodes. Then, the contaminants that accumulate at the electrodes are treated and eliminated with various physical–chemical approaches (electroplating, precipitation, pump-and-treat the water near the electrodes, or sorption with ion-exchange resins).

The electrokinetic method is used since it is efficacious, even in low permeability soils, and does not excessively change soil properties [86]. The combination of extraction with low-molecular-weight organic acids and electrochemical adsorption promotes the decrease of pollutant concentrations in soil [87–89]. Different acids can be used to extract heavy metals from the soil: sulfuric, ethylene diamine tetraacetic (EDTA), acetic, and citric acid. Cameselle and Pena [86] demonstrated that the use of citric acid efficiently removes cadmium, cobalt, chromium, copper, lead, and zinc, since it favors the acidification of soil, solubilization of metals, their transportation by electro-osmosis, and their electromigration towards the cathode. The efficiency of citric acid is also relevant in the removal of arsenic [90].

Soil acidification is not environmentally acceptable, and this must be assessed before the choice of technique that will be adopted is made. A low pH has two severe drawbacks:

- In these conditions, usually, the metals are in an ionic form; this is to say they can be mobilized from soil.
- The vegetal regeneration of soil could be limited and made more difficult.

Electrochemical remediation is rather efficient, especially in saturated clay soils, but it is rather complex to carry out.

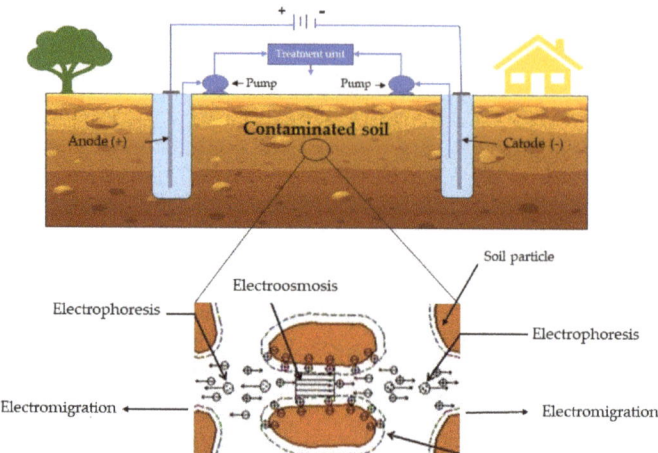

Figure 4. Electrochemical remediation scheme.

4.3. Biological Processes

Biological processes permit the removal of heavy metals and metalloids by exploiting microorganisms and plants. Biological agents respond to contamination through their defense mechanisms, such as enzyme secretion and cellular morphological changes. Typical biological techniques are bioaccumulation, bioleaching, biosorption, phytostabilization, and phytoextraction (as showed in Figure 5). Heavy metals and metalloids are not degraded by microorganisms or plants, but they are accumulated, stabilized, or bonded in less toxic volatile compounds. Specific microbial strains can encourage the degradation of pollutants and also improve soil characteristics, such as fertility [93].

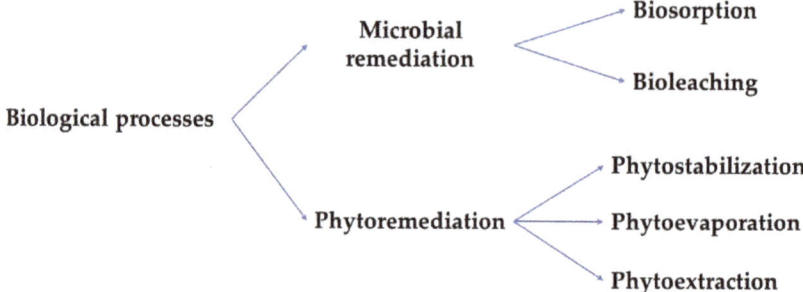

Figure 5. Biological processes to remove heavy metals and metalloids.

4.3.1. Microbial Bioremediation

The removal of heavy metals and metalloids can be actuated by fostering the growth and development of specific microorganisms in contaminated soil, namely through bioaugmentation and/or biostimulation processes. Heavy metals/metalloids can be oxidized, reduced, immobilized, and metabolized by microorganisms. Different from organic pollutants, heavy metals/metalloids are not degraded by microorganisms, but simply accumulate and are adsorbed at the binding sites present in the cellular structure, for example into microbial tissue.

The main microbial processes are biosorption and bioleaching, and are described in the study by Zabochnicka-Świątek and Krzywonos [135].

Biosorption

In the biosorption process, the heavy metals are immobilized onto the cellular structure of microorganisms. This is possible through the extracellular binding created between the cell surface (anions) and metal ions (cations). Extracellular materials have active functional groups that promote the binding mechanisms. The biosorption mechanism is complex and may include physical adsorption, ion exchange, complex formation, reduction, and precipitation. The efficiency of biosorption is influenced by several factors: metal ion properties, process conditions, density of sorption centers, and types of immobilization agents, as shown in the study by Velkova et al. [136].

Typical bioadsorbents are bacteria [91–94], fungi [95–98], and algae [99]:

- Bacteria: heavy metal ions can be bound and accumulated on polysaccharide slime layers of bacteria through functional groups, such as carboxyl, amino, phosphate, or sulfate groups.
- Fungi: are used to adsorb the heavy metals and metalloids through ion exchange and coordination in the chitin–chitosan complex, glucuronic acid, phosphate, and polysaccharides present in their cells.
- Algae: the absorption of heavy metals/metalloids occurs since algae form peptides as a defense mechanism. The functional groups (carboxyl, amino, sulfhydryl, and sulfonate) are among the constituents of the algal cell wall, and ion exchange promotes the adsorption of metal ions.

Bioleaching

The bioleaching process reduces mobility and stabilizes pollutants using the capacity of microorganisms to produces secretions, such as low molecular weight organic acids, that can dissolve heavy metals and soil particles containing heavy metal ores. In this way, heavy metals are directly solubilized by the metabolism of microorganisms or indirectly by their metabolites [101]. The agents that promote the leaching are biosurfactants, namely polysaccharides, lipids, and lipopeptides, produced by microorganisms, with a high surface activity that permits the formation of chelating metals and binding of metal ions [100]. This method has been extensively studied since it is an environmentally-friendly and inexpensive technique. For example, Yang et al. [100] showed the capability of a biosurfactant produced by a microbial strain isolated from cafeteria sewer sludge.

In general, all microbial biodegradation is low cost and eco-friendly. However, there are some limits:

- several microorganisms cannot bind toxic metals into harmless metabolites;
- the process is slower than others;
- it is efficient if the environmental conditions are suitable for microbial metabolism.

4.3.2. Phytoremediation

Phytoremediation is a bioremediation technique that exploits the capacity of plants to intercept, take up, accumulate, adsorb, or stabilize contaminants. The phytoremediation process aims to sequester contaminants via the roots of plants to lesser toxic elements or absorb them to the roots or shoots [102]. Some plants, called hyperaccumulators, can tolerate and accumulate more than 1000 mg·kg^{-1} of dry matter of copper, cadmium, chromium, lead, nickel, cobalt, or up to 10,000 mg·kg^{-1} of dry matter; this is to say, 1% by weight on a dry basis, of zinc or manganese [137].

The process is slow, and, for this reason, if a plant has large roots and/or shoots, large quantities of pollutants can be accumulated.

Several trees, as well as agricultural and herbaceous crops, are used, such as willow (*Salix* spp.) [103–105], poplar (*Populus* spp.) [102–104], wheat (*Triticumaestivum* L.) [106], sweet and grain sorghum (*Sorghum bicolor*) [107,108], and smilo grass (*Piptatherum miliaceum*) [109]. The phytoremediation process can be assisted by natural chelators, which promote the bioavailability and adsorption of heavy metals, such as copper [138].

Depending on the type of pollutants, plants, and removal mechanism, the phytoremediation techniques can be divided into phytostabilization, phytoevaporation, and phytoextraction.

Phytostabilization

In phytostabilization, plants prevent the mobility and the bioaccumulation of heavy metals and metalloids. The mechanism entails the complexation of metal ions with the roots, or with the cell walls, binding with molecules present in the roots, such as phytochelatins and metallothioneins, and finally sequestering them to root vacuoles [139]. Saran et al. [111] studied *Helianthus petiolaris* to treat soil contaminated with cadmium and lead, and showed that this plant does not interact with the activity of microorganisms already present in the root zone, and is efficient if the soil Cd concentration is lower than 50 mg·kg^{-1} and the Pb concentration is lower than 1000 mg·kg^{-1}.

Phytoevaporation

In phytoevaporation, heavy metals and metalloids are bound to volatile organic compounds which are then released into the atmosphere through the transpiration process of the plants. Sakakibara et al. [117] studied the phytovolatilization of arsenic from soil using *Pteris vittata* plant. This plant is capable of metabolizing arsenic into volatile forms (arsenite and arsenate), reducing the arsenic concentration in soil. In nature, few plants are able to carry out phytoevaporation. For this reason, genetically modified plants are used, such as transgenic tobacco plants, that are able to remove mercury from contaminated soils [116]. This technique is not exempt from drawbacks. The main problem is that the volatile substances emitted into the atmosphere can be still toxic, and their emission constitutes secondary pollution around the plants themselves.

Phytoextraction

Phytoextraction is an absorption process of heavy metals and metalloids from soil through the plant roots. Then, the pollutants are transported and accumulated in the aboveground biomass, such as shoots. Bioaccumulation in shoots is advantageous for harvest. Mahmood-ul-Hassan et al. [112] studied different non-eatable floriculture plants, i.e., antirrhinum, pansy, calendula, and marigold, in soil contaminated by cadmium, chromium, nickel, and lead. They added a bacterial inoculum and EDTA, finding that both promoted the growth of biomass and the accumulation of contaminants in the roots, and this limited the transfer of contaminants to the shoots.

Phytoremediation allows to have a cheap and ecologically sustainable approach, but it is slow and depends on plant growth, seasonality, and environmental conditions. For these reasons, phytoremediation is often combined with other removal techniques. Combination is useful for improving removal efficiency, for example, using:

- phytoremediation + bioaugmentation;
- phytoremediation + EDTA amended soil [112];
- electrokinetic remediation + phytoremediation [113].

Agnello et al. [114] compared biological processes such as natural attenuation, bioaugmentation, and phytoremediation. They found that bioaugmentation-assisted phytoremediation optimizes the synergy effect of plants and microorganisms, improving the removal of heavy metals and metalloids [115].

5. Case Studies and Estimation Costs

At a large scale, the choice of technique for the removal of heavy metals and metalloids must consider several factors: removal efficiency, cost, treatment duration, pollutant concentration, and soil properties. In the literature, there are few case studies and field applications of various techniques for heavy metal remediation. Some examples are summarized in Table 5.

Table 5. Case studies for heavy metals/metalloids.

Technique	Metals/Metalloids	Size	Location	Year	Ref.
Soil replacement	Hg, Cd, Ni, Cu, Cr, Pb, Zn, As, Sb, Ba, Be, Mo, Se	-	Serbia	2014	[140]
Electrokinetic remediation	As, Cu, Pb	26.25 m^3	Janghang, South Korea	2013	[141]
Electrokinetic remediation	-	57 m^2	Paducah, Kentucky	1997	[142]
Chemical stabilization	Cd, Pb, Zn	10 m^2	Biscay, Spain	2012	[143]
Soil washing and phytoremediation	Cd, Zn, Pb, Cu	64 m^2 (8 plots of 8 m^2)	Shaoguan, China	2011	[144]
Phytoremediation	Cr, Zn, As, Cd, Pb	1600 m^2	Taranto, Italy	2013	[102]
Phytoremediation	As, Zn, Pb, Cd	1000 m^2	Porto Marghera, Italy	2008	[145]

Several companies are specialized in the stabilization/solidification technique at an industrial scale, but they do not usually publish financial information.

It is difficult to estimate actual costs due to a lack of recent and large-scale data. For example, by referencing 2010, the cost for an electrokinetic method that implements a 1 V·cm^{-1} voltage slope to eliminate copper is about 13–40 USD·m^{-3} of soil [89].

Chen and Li [146] had estimated the costs for soil washing and phytoextraction (year 2000):

- for soil washing, the cost is in the range of 200 USD·m^{-3} (small sites) to 70 USD·m^{-3} (large sites);
- for phytoextraction, the cost can go from 35 USD·m^{-3} (small sites) to 10 USD·m^{-3} (large sites).

6. Conclusions

Around the world, many polluted sites have been and are currently being cleaned up by adopting one of the above-described techniques. However, as noted, little data on large-scale operations can be found in the literature. A desirable outcome is that the shortage of data on remediation at a real-scale is filled, and is supported with tighter cooperation between research and industry. This will be of help to drive future soil clean-up, which is sometimes urgent and cannot wait for experimental findings to be achieved using new techniques.

Author Contributions: Writing—original draft preparation, C.M.R.; writing—review and editing, C.M.R., F.C., S.S.; supervision, F.C. All authors have read and agreed to the published version of the manuscript.

Funding: This research received no external funding.

Institutional Review Board Statement: Not applicable.

Informed Consent Statement: Not applicable.

Data Availability Statement: Not applicable.

Conflicts of Interest: The authors declare no conflict of interest.

References

1. Atieh, M.A.; Ji, Y.; Kochkodan, V. Metals in the Environment: Toxic Metals Removal. *Bioinorg. Chem. Appl.* **2017**, *2017*, 4309198. [CrossRef]
2. He, Z.; Shentu, J.; Yang, X.; Baligar, V.C.; Zhang, T.; Stoffella, P.J. Heavy Metal Contamination of Soil: Sources, Indicators, and Assessment. *J. Environ. Indic.* **2015**, *9*, 17–18.
3. Khalid, S.; Shahid, M.; Niazi, N.K.; Murtaza, B.; Bibi, I.; Dumat, C. A comparison of technologies for remediation of heavy metal contaminated soils. *J. Geochem. Explor.* **2017**, *182*, 247–268. [CrossRef]
4. Cheng, S.; Chen, T.; Xu, W.; Huang, J.; Jiang, S.; Yan, B. Application Research of Biochar for the Remediation of Soil Heavy Metals Contamination: A Review. *Molecules* **2020**, *25*, 3167. [CrossRef] [PubMed]
5. Tóth, G.; Hermann, T.; Da Silva, M.R.; Montanarella, L. Heavy metals in agricultural soils of the European Union with implications for food safety. *Environ. Int.* **2016**, *88*, 299–309. [CrossRef] [PubMed]

6. Li, L.; Zhang, Y.; Ippolito, J.A.; Xing, W.; Qiu, K.; Yang, H. Lead smelting effects heavy metal concentrations in soils, wheat, and potentially humans. *Environ. Pollut.* **2020**, *257*, 113614. [CrossRef] [PubMed]
7. Yu, F.; Lin, J.; Xie, D.; Yao, Y.; Wang, X.; Huang, X.; Xin, M.; Yang, F.; Liu, K.; Li, Y. Soil properties and heavy metal concentrations affect the composition and diversity of the diazotrophs communities associated with different land use types in a mining area. *Appl. Soil Ecol.* **2020**, *155*, 1030669. [CrossRef]
8. Yan, D.; Bai, Z.; Liu, X. Heavy-Metal Pollution Characteristics and Influencing Factors in Agricultural Soils: Evidence from Shuozhou City, Shanxi Province, China. *Sustainability* **2020**, *12*, 1907. [CrossRef]
9. Bocardi, J.M.B.; Pletsch, A.L.; Melo, V.F.; Quinaina, S.P. Quality reference values for heavy metals in soils developed from basic rocks under tropical conditions. *J. Geochem. Explor.* **2020**, *217*, 106591. [CrossRef]
10. Brito, A.C.C.; Boechat, C.L.; de Sena, A.F.S.; Duarte, L.d.S.L.; do Nascimento, C.W.A.; da Silva, Y.J.A.B.; Saraiva, P.C. Assessing the Distribution and Concentration of Heavy Metals in Soils of an Agricultural Frontier in the Brazilian Cerrado. *Water Air Soil Pollut.* **2020**, *231*, 388. [CrossRef]
11. Dhaliwal, S.S.; Singh, J.; Taneja, P.K.; Mandal, A. Remediation techniques for removal of heavy metals from the soil contaminated through different sources: A review. *Environ. Sci. Pollut. Res.* **2020**, *27*, 1319–1333. [CrossRef]
12. Sivakumar, D.; Kandaswamy, A.N.; Gomathi, V.; Rajeshwaran, R.; Murugan, N. Bioremediation studies on reduction of heavy metals toxicity. *Pollut. Res.* **2014**, *33*, 553–558.
13. Ali, H.; Khan, E. What are heavy metals? Long-standing controversy over the scientific use of the term 'heavy metals'—Proposal of a comprehensive definition. *Toxicol. Environ. Chem.* **2018**, *100*, 6–19. [CrossRef]
14. Caporale, A.G.; Violante, A. Chemical Processes Affecting the Mobility of Heavy Metals and Metalloids in Soil Environments. *Curr. Pollut. Rep.* **2016**, *2*, 15–27. [CrossRef]
15. Ali, H.; Khan, E.; Ilahi, I. Environmental Chemistry and Ecotoxicology of Hazardous Heavy Metals: Environmental Persistence, Toxicity, and Bioaccumulation. *J. Chem.* **2019**, *2019*, 6730305. [CrossRef]
16. Salomons, W.; Forstner, U.; Mader, P. *Heavy Metals Problems and Solutions*, 1st ed.; Springer: Heidelberg, Germany, 1995; ISBN 9783642793189.
17. Koller, M.; Saleh, H.M. Introductory Chapter: Introducing Heavy Metals. In *Heavy Metals*; Saleh, H.M., Aglan, R., Eds.; IntechOpen: London, UK, 2018.
18. Wang, L.K.; Chen, J.P.; Hung, Y.-T.; Shammas, N.K. *Heavy Metals in the Environment*; Taylor & Francis Group: Broken Sound Parkway, NW, USA, 2009; ISBN 9781420073164.
19. Statista. Available online: https://www.statista.com/ (accessed on 2 December 2020).
20. Kinuthia, G.K.; Ngure, V.; Beti, D.; Lugalia, R.; Wangila, A.; Kamau, L. Levels of heavy metals in wastewater and soil samples from open drainage channels in Nairobi, Kenya: Community health implication. *Sci. Rep.* **2020**, *10*, 8434. [CrossRef] [PubMed]
21. New York State. *New York State Brownfield Cleanup Program Development of Soil Cleanup Objectives Technical Support Document*; New York State Department of Environmental Conservation and New York State Department of Health: Albany, NY, USA, 2006.
22. Decreto Legislativo n. 152 del 3 aprile 2006 "Norme in materia ambientale", Supplemento Ordinario alla "Gazzetta Ufficiale" n. 88 del 14 aprile 2006. Available online: https://www.camera.it/parlam/leggi/deleghe/06152dl.htm (accessed on 2 December 2020).
23. *Government Decree on the Assessment of Soil Contamination and Remediation Needs 214/2007, 1 March 2007*; Ministry of Environment: Helsinki, Finland, 2007; (the legally binding document is In Finnish or Swedish).
24. *Soil, Ground Water and Sediment Standards for Use under Part XV.1 of the Environmental Protection Act*; Canadian Ministry of the Environment (CME): Toronto, ON, Canada, 15 April 2011.
25. Real Decreto 1310/1990, de 29 de octubre, por el que se regula la utilización de los lodos de depuración en el sector agrario. Available online: https://www.boe.es/buscar/doc.php?id=BOE-A-1990-26490 (accessed on 2 December 2020).
26. Zhou, W.; Han, G.; Liu, M.; Song, C.; Li, X.; Malem, F. Vertical Distribution and Controlling Factors Exploration of Sc, V, Co, Ni, Mo and Ba in Six Soil Profiles of The Mun River Basin, Northeast Thailand. *Int. J. Environ. Res. Public Health* **2020**, *17*, 1745. [CrossRef]
27. Vacca, A.; Bianco, M.R.; Murolo, M.; Violante, P. Heavy metals in contaminated soils of the Rio Sitzerri floodplain (Sardinia, Italy): Characterization and impact on pedodiversity. *Land Degrad. Dev.* **2012**, *23*, 350–364. [CrossRef]
28. Cicchella, D.; Giaccio, L.; Lima, A.; Albanese, S.; Cosenza, A.; Civitillo, D.; De Vivo, B. Assessment of the topsoil heavy metals pollution in the Sarno River basin, south Italy. *Environ. Earth Sci.* **2014**, *71*, 5129–5143. [CrossRef]
29. Marrugo-Negrete, J.; Pinedo-Hernández, J.; Díez, S. Assessment of heavy metal pollution, spatial distribution and origin in agricultural soils along the Sinú River Basin, Colombia. *Environ. Res.* **2017**, *154*, 380–388. [CrossRef]
30. Fan, T.T.; Wang, Y.-J.; Li, C.-B.; He, J.-Z.; Gao, J.; Zhou, D.-M.; Friedman, S.P.; Sparks, D.L. Effect of Organic Matter on Sorption of Zn on Soil: Elucidation by Wien Effect Measurements and EXAFS Spectroscopy. *Environ. Sci. Technol.* **2016**, *50*, 2931–2937. [CrossRef] [PubMed]
31. Shen, F.; Liao, R.; Ali, A.; Mahar, A.; Guo, D.; Li, R.; Sun, X.; Awasthi, M.K.; Wang, Q.; Zhang, Z. Spatial distribution and risk assessment of heavy metals in soil near a Pb/Zn smelter in Feng County, China. *Ecotoxicol. Environ. Saf.* **2017**, *139*, 254–262. [CrossRef] [PubMed]
32. Jiang, Y.; Chao, S.; Liu, J.; Yang, Y.; Chen, Y.; Zhang, A.; Cao, H. Source apportionment and health risk assessment of heavy metals in soil for a township in Jiangsu Province, China. *Chemosphere* **2017**, *168*, 1658–1668. [CrossRef] [PubMed]

33. Jamal, A.; Sarim, M. Heavy metals distribution in different soil series of district Swabi, Khyber Pakhunkhawa, Pakistan. *World Sci. News* **2018**, *105*, 1–13.
34. Kelepertzis, E. Accumulation of heavy metals in agricultural soils of Mediterranean: Insights from Argolida basin, Peloponnese, Greece. *Geoderma* **2014**, *221–222*, 82–90. [CrossRef]
35. Hejna, M.; Gottardo, D.; Baldi, A.; Dell'Orto, V.; Cheli, F.; Zaninelli, M.; Rossi, L. Review: Nutritional ecology of heavy metals. *Animal* **2018**, *12*, 2156–2170. [CrossRef]
36. Mahurpawar, M. Effects of heavy metals on human health. *Int. J. Res. Granthaalayah* **2015**, *3*, 1–7. [CrossRef]
37. Jan, A.T.; Azam, M.; Siddiqui, K.; Ali, A.; Choi, I.; Haq, Q.M.R. Heavy Metals and Human Health: Mechanistic Insight into Toxicity and Counter Defense System of Antioxidants. *Int. J. Mol. Sci.* **2015**, *16*, 29592–29630. [CrossRef]
38. Dixit, R.; Wasiullah; Malaviya, D.; Pandiyan, K.; Singh, U.B.; Sahu, A.; Shukla, R.; Singh, B.P.; Rai, J.P.; Sharma, P.K.; et al. Bioremediation of Heavy Metals from Soil and Aquatic Environment: An Overview of Principles and Criteria of Fundamental Processes. *Sustainability* **2015**, *7*, 2189–2212. [CrossRef]
39. Shekhawat, K.; Chatterjee, S.; Joshi, B. Chromium Toxicity and its Health Hazards. *Int. J. Adv. Res.* **2015**, *7*, 167–172.
40. Mishra, S.; Bharagava, R.N. Toxic and genotoxic effects of hexavalent chromium in environment and its bioremediation strategies. *J. Environ. Sci. Heal. Part C* **2016**, *34*, 1–32. [CrossRef]
41. Kumar, V.; Kalita, J.; Misra, U.K.; Bora, H.K. A study of dose response and organ susceptibility of copper toxicity in a rat model. *J. Trace Elem. Med. Biol.* **2015**, *29*, 269–274. [CrossRef] [PubMed]
42. Hao, X.; Xie, P.; Zhu, Y.; Taghavi, S.; Wei, G.; Rensing, C. Copper Tolerance Mechanisms of Mesorhizobium amorphae and Its Role in Aiding Phytostabilization by Robinia pseudoacacia in Copper Contaminated Soil. *Environ. Sci. Technol.* **2015**, *49*, 2328–2340. [CrossRef] [PubMed]
43. Saleem, M.H.; Ali, S.; Kamran, M.; Iqbal, N.; Azeem, M.; Javed, M.T.; Ali, Q.; Haider, M.Z.; Irshad, S.; Rizwan, M.; et al. Ethylenediaminetetraacetic Acid (EDTA) Mitigates the Toxic Effect of Excessive Copper Concentrations on Growth, Gaseous Exchange and Chloroplast Ultrastructure of Corchorus capsularis L. and Improves Copper Accumulation Capabilities. *Plants* **2020**, *9*, 756. [CrossRef] [PubMed]
44. Krężel, A.; Maret, W. The biological inorganic chemistry of zinc ions. *Arch. Biochem. Biophys.* **2016**, *611*, 3–19. [CrossRef] [PubMed]
45. Alkan, F.A.; Kilinc, E.; Gulyasar, T.; Or, M.E. Evaluation of zinc and copper toxicity caused by ingestion of Turkish coins: An in vitro study. *J. Elementol.* **2020**, *25*, 961–971. [CrossRef]
46. Ackova, D.G. Heavy metals and their general toxicity on plants. *Plant Sci. Today* **2018**, *5*, 15–19.
47. Rahimzadeh, M.R.; Rahimzadeh, M.R.; Kazemi, S.; Moghadamnia, A. Cadmium toxicity and treatment: An update. *Casp. J. Intern. Med.* **2017**, *8*, 135–145. [CrossRef]
48. Liu, L.; Shank, Y.K.; Li, L.; Chen, Y.H.; Qin, Z.Z.; Zhou, L.J.; Yuan, M.; Ding, C.B.; Liu, J.; Huang, Y.; et al. Cadmium stress in Dongying wild soybean seedling: Growth, Cd accumulation, and photosynthesis. *Photosynthetica* **2018**, *56*, 1346–1352. [CrossRef]
49. Wani, A.L.; Ara, A.; Usmani, J.A. Lead toxicity: A review. *Interdiscip. Toxicol.* **2015**, *8*, 55–64. [CrossRef]
50. Gidlow, D.A. Lead toxicity. *Occup. Med.* **2015**, *65*, 348–356. [CrossRef] [PubMed]
51. Andreoli, V.; Sprovieri, F. Genetic Aspects of Susceptibility to Mercury Toxicity: An Overview. *Int. J. Environ. Res. Public Health* **2017**, *14*, 93. [CrossRef]
52. Park, J.-D.; Zheng, W. Human Exposure and Health Effects on Inorganic and Elemental Mecury. *J. Prev. Med. Public Health* **2012**, *45*, 344–352. [CrossRef] [PubMed]
53. Farooq, M.A.; Islam, F.; Ali, B.; Najeeb, U.; Mao, B.; Gill, R.A.; Yan, G.; Siddique, K.H.M.; Zhou, W. Arsenic toxicity in plants: Cellular and molecular mechanisms of its transport and metabolism. *Environ. Exp. Bot.* **2016**, *132*, 42–52. [CrossRef]
54. Awasthi, S.; Chauhan, R.; Srivastava, S.; Tripathi, R.D.; Peña-Castro, J.M. The Journey of Arsenic from Soil to Grain in Rice. *Front. Plant Sci.* **2017**, *8*, 1–13. [CrossRef]
55. Shrivastava, A.; Ghosh, D.; Dash, A.; Bose, S. Arsenic Contamination in Soil and Sediment in India: Sources, Effects, and Remediation. *Curr. Pollut. Rep.* **2015**, *1*, 35–56. [CrossRef]
56. Hàkanson, L. An Ecological Risk Index for Aquatic Pollution Control. A Sedimentological Approach. *Water Res.* **1980**, *14*, 975–1001. [CrossRef]
57. Pan, X.-D.; Wu, P.-G.; Jiang, X.-G. Levels and potential health risk of heavy metals in marketed vegetables in Zhejiang, China. *Sci. Rep.* **2016**, *6*, 20317. [CrossRef]
58. Baran, A.; Wieczorek, J.; Mazurek, R.; Urbański, K.; Klimkowicz-Pawlas, A. Potential ecological risk assessment and predicting zinc accumulation in soils. *Environ. Geochem. Health* **2018**, *40*, 435–450. [CrossRef]
59. Wu, Q.; Leung, J.Y.S.; Geng, X.; Chen, S.; Huang, X.; Li, H.; Huang, Z.; Zhu, L.; Chen, J.; Lu, Y. Heavy metal contamination of soil and water in the vicinity of an abandoned e-waste recycling site: Implications for dissemination of heavy metals. *Sci. Total Environ.* **2015**, *506–507*, 217–225. [CrossRef]
60. Song, B.; Zeng, G.; Gong, J.; Liang, J.; Xu, P.; Liu, Z.; Zhang, Y.; Zhang, C.; Cheng, M.; Liu, Y.; et al. Evaluation methods for assessing effectiveness of in situ remediation of soil and sediment contaminated with organic pollutants and heavy metals. *Environ. Int.* **2017**, *105*, 43–55. [CrossRef]
61. Gong, Y.; Zhao, D.; Wang, Q. An overview of field-scale studies on remediation of soil contaminated with heavy metals and metalloids: Technical progress over the last decade. *Water Res.* **2018**, *147*, 440–460. [CrossRef] [PubMed]

62. Ashraf, S.; Ali, Q.; Zahir, Z.A.; Ashraf, S.; Asghar, H.N. Phytoremediation: Environmentally sustainable way for reclamation of heavy metal polluted soils. *Ecotoxicol. Environ. Saf.* **2019**, *174*, 714–727. [CrossRef]
63. Boente, C.; Sierra, C.; Rodrìguez-Valdés, E.; Menéndez-Aguado, J.M.; Gallego, J.R. Soil washing optimization by means of attributive analysis: Case study for the removal of potentially toxic elements from soil contaminated with pyrite ash. *J. Clean. Prod.* **2017**, *142*, 2693–2699. [CrossRef]
64. Liao, X.; Li, Y.; Yan, X. Removal of heavy metals and arsenic from a co-contaminated soil by sieving combined with washing process. *J. Environ. Sci.* **2016**, *41*, 202–210. [CrossRef] [PubMed]
65. Park, B.; Son, Y. Ultrasonic and mechanical soil washing processes for the removal of heavy metals from soils. *Ultrason. Sonochem.* **2017**, *35*, 640–645. [CrossRef] [PubMed]
66. Yi, Y.M.; Sung, K. Influence of washing treatment on the qualities of heavy metal–contaminated soil. *Ecol. Eng.* **2015**, *81*, 89–92. [CrossRef]
67. Wang, P.; Hu, X.; He, Q.; Waigi, M.G.; Wang, J.; Ling, W. Using Calcination Remediation to Stabilize Heavy Metals and Simultaneously Remove Polycyclic Aromatic Hydrocarbons in Soil. *Int. J. Environ. Res. Public Health* **2018**, *15*, 1731. [CrossRef] [PubMed]
68. Chang, T.C.; Yen, J.H. On-site mercury-contaminated soils remediation by using thermal desorption technology. *J. Hazard. Mater.* **2006**, *128*, 208–217. [CrossRef]
69. Ballesteros, S.; Rincón, J.M.; Rincón-Mora, B.; Jordán, M.M. Vitrification of urban soil contamination by hexavalent chromium. *J. Geochem. Explor.* **2017**, *174*, 132–139. [CrossRef]
70. Ullah, A.; Ma, Y.; Li, J.; Tahir, N.; Hussain, B. Effective Amendments on Cadmium, Arsenic, Chromium and Lead Contaminated Paddy Soil for Rice Safety. *Agronomy* **2020**, *10*, 359. [CrossRef]
71. Islam, M.N.; Taki, G.; Nguye, X.P.; Jo, Y.-T.; Kim, J.; Park, J.-H. Heavy metal stabilization in contaminated soil by treatment with calcined cockle shell. *Environ. Sci. Pollut. Res.* **2017**, *24*, 7177–7183. [CrossRef]
72. Shi, L.; Zhang, X.; Chen, Z. Removal of Chromium (VI) from wastewater using bentonite-supported nanoscale zero-valent iron. *Water Res.* **2011**, *45*, 886–892. [CrossRef] [PubMed]
73. Zhang, M.; Yi, K.; Zhang, X.; Han, P.; Liu, W.; Tong, M. Modification of zero valent iron nanoparticles by sodium alginate and bentonite: Enhanced transport, effective hexavalent chromium removal and reduced bacterial toxicity. *J. Hazard. Mater.* **2020**, *388*, 121822. [CrossRef]
74. Li, Z.; Wang, L.; Wu, J.; Xu, Y.; Wang, F.; Tang, X.; Xu, J.; Ok, Y.S.; Meng, J.; Liu, X. Zeolite-supported nanoscale zero-valent iron for immobilization of cadmium, lead, and arsenic in farmland soils: Encapsulation mechanisms and indigenous microbial responses. *Environ. Pollut.* **2020**, *260*, 114098. [CrossRef] [PubMed]
75. Li, Z.; Wang, L.; Meng, J.; Liu, X.; Xu, J.; Wang, F.; Brookes, P. Zeolite-supported nanoscale zero-valent iron: New findings on simultaneous adsorption of Cd(II), Pb(II), and As(III) in aqueous solution and soil. *J. Hazard. Mater.* **2018**, *344*, 1–11. [CrossRef] [PubMed]
76. Fan, J.; Chen, X.; Xu, Z.; Xu, X.; Zhao, L.; Qui, H.; Cao, X. One-pot synthesis of nZVI-embedded biochar for remediation of two mining arsenic-contaminated soils: Arsenic immobilization associated with iron transformation. *J. Hazard. Mater.* **2020**, *398*, 122901. [CrossRef] [PubMed]
77. Wang, H.; Zhang, M.; Li, H. Synthesis of Nanoscale Zerovalent Iron (nZVI) supported on biochar for chromium remediation from aqueous solution and soil. *Int. J. Environ. Res. Public Health* **2019**, *16*, 4430. [CrossRef] [PubMed]
78. Pei, G.; Zhu, Y.; Wen, J.; Pei, Y.; Li, H. Vinegar residue supported nanoscale zero-valent iron: Remediation of hexavalent chromium in soil. *Environ. Pollut.* **2020**, *256*, 113407. [CrossRef] [PubMed]
79. Xue, W.; Peng, Z.; Huang, D.; Zeng, G.; Wan, J.; Xu, R.; Cheng, M.; Zhang, C.; Jiang, D.; Hu, Z. Nanoremediation of cadmium contaminated river sediments: Microbial response and organic carbon changes. *J. Hazard. Mater.* **2018**, *359*, 290–299. [CrossRef]
80. Tomašević Filipović, D.; Kerkez, D.; Dalmacija, B.; Slijepčević, N.; Krčmar, D.; Rađenović, D.; Bečelić-Tomin, M. Remediation of toxic metal contaminated sediment using three types of nZVI supported materials. *Bull. Environ. Contam. Toxicol.* **2018**, *101*, 725–731. [CrossRef] [PubMed]
81. Sun, Y.; Zheng, F.; Wang, W.; Zhang, S.; Wang, F. Remediation of Cr(VI)-contaminated soil by Nano-Zero-Valent Iron in combination with biochar or humic acid and the consequences for plant performance. *Toxics* **2020**, *8*, 26. [CrossRef] [PubMed]
82. Liu, S.-J.; Jiang, J.-Y.; Wang, S.; Guo, Y.-P.; Ding, H. Assessment of water-soluble thiourea-formaldehyde (WTF) resin for stabilization/solidification (S/S) of heavy metal contaminated soils. *J. Hazard. Mater.* **2018**, *346*, 167–173. [CrossRef] [PubMed]
83. Feng, W.; Zhang, S.; Zhong, Q.; Wang, G.; Pan, X.; Xu, X.; Zhou, W.; Li, T.; Luo, L.; Zhang, Y. Soil washing remediation of heavy metal from contaminated soil with EDTMP and PAA: Properties, optimization, and risk assessment. *J. Hazard. Mater.* **2020**, *381*, 120997. [CrossRef]
84. Wang, G.; Pan, X.; Zhang, S.; Zhong, Q.; Zhou, W.; Zhang, X.; Wu, J.; Vijver, M.G.; Peijnenburg, W.J.G.M. Remediation of heavy metal contaminated soil by biodegradable chelator–induced washing: Efficiencies and mechanisms. *Environ. Res.* **2020**, *186*, 109554. [CrossRef]
85. Zhai, X.; Li, Z.; Huang, B.; Luo, N.; Huang, M.; Zhang, Q.; Zeng, G. Remediation of multiple heavy metal-contaminated soil through the combination of soil washing and in situ immobilization. *Sci. Total Environ.* **2018**, *635*, 92–99. [CrossRef]
86. Cameselle, C.; Pena, A. Enhanced electromigration and electro-osmosis for the remediation of an agricultural soil contaminated with multiple heavy metals. *Process Saf. Environ. Prot.* **2016**, *104*, 209–217. [CrossRef]

87. Yang, X.; Liu, L.; Tan, W.; Liu, C.; Dang, Z.; Qiu, G. Remediation of heavy metal contaminated soils by organic acid extraction and electrochemical adsorption. *Environ. Pollut.* **2020**, *264*, 114745. [CrossRef]
88. Xu, J.; Liu, C.; Hsu, P.-C.; Zhao, J.; Wu, T.; Tang, J.; Liu, K.; Cui, Y. Remediation of heavy metal contaminated soil by asymmetrical alternating current electrochemistry. *Nat. Commun.* **2019**, *10*, 2240. [CrossRef]
89. Han, J.-G.; Hong, K.-K.; Kim, Y.-W.; Lee, J.-Y. Enhanced electrokinetic (E/K) remediation on copper contaminated soil by CFW (carbonized foods waste). *J. Hazard. Mater.* **2010**, *177*, 530–538. [CrossRef]
90. Rosa, M.A.; Egido, J.A.; Márquez, M.C. Enhanced electrochemical removal of arsenic and heavy metals from mine tailings. *J. Taiwan Inst. Chem. Eng.* **2017**, *78*, 409–415. [CrossRef]
91. Jaafer, R.; Al-Sulami, A.; Al-Taee, A. The biosorption ability of shewanella oneidensis for cadmium and lead isolated from soil in Basra Governorate, Iraq. *Pollut. Res.* **2019**, *38*, 267–270.
92. Akhter, K.; Ghous, T.; Andleeb, S.; Nasim, F.H.; Ejaz, S.; Zain-ul-Abdin; Khan, B.A.; Ahmed, M.N. Bioaccumulation of Heavy Metals by Metal-Resistant Bacteria Isolated from Tagetes minuta Rhizosphere, Growing in Soil Adjoining Automobile Workshops. *Pak. J. Zool.* **2017**, *49*, 1841–1846. [CrossRef]
93. Cui, Z.; Zhang, X.; Yang, H.; Sun, L. Bioremediation of heavy metal pollution utilizing composite microbial agent of Mucor circinelloides, Actinomucor sp. and Mortierella sp. *J. Environ. Chem. Eng.* **2017**, *5*, 3616–3621. [CrossRef]
94. Wang, T.; Sun, H.; Ren, X.; Li, B.; Mao, H. Evaluation of biochars from different stock materials as carriers of bacterial strain for remediation of heavy metal-contaminated soil. *Sci. Rep.* **2017**, *7*, 12114. [CrossRef] [PubMed]
95. Bano, A.; Hussain, J.; Akbar, A.; Mehmood, K.; Anwar, M.; Hasni, M.S.; Ullah, S.; Sajid, S.; Ali, I. Biosorption of heavy metals by obligate halophilic fungi. *Chemosphere* **2018**, *199*, 218–222. [CrossRef]
96. Hassan, A.; Pariatamby, A.; Ahmed, A.; Auta, H.S.; Hamid, F.S. Enhanced Bioremediation of Heavy Metal Contaminated Landfill Soil Using Filamentous Fungi Consortia: A Demonstration of Bioaugmentation Potential. *Water Air Soil Pollut.* **2019**, *230*. [CrossRef]
97. Hassan, A.; Periathamby, A.; Ahmed, A.; Innocent, O.; Hamid, F.S. Effective bioremediation of heavy metal–contaminated landfill soil through bioaugmentation using consortia of fungi. *J. Soils Sediments* **2020**, *20*, 66–80. [CrossRef]
98. Iram, S.; Shabbir, R.; Zafar, H.; Javaid, M. Biosorption and Bioaccumulation of Copper and Lead by Heavy Metal-Resistant Fungal Isolates. *Arab. J. Sci. Eng.* **2015**, *40*, 1867–1873. [CrossRef]
99. Taleel, M.M.; Ghoml, N.K.; Jozl, S.A. Arsenic Removal of Continared Soil by Phytoremediation of Vetiver Grass, Chara Algae and Water Hyacinth. *Bull. Environ. Contam. Toxicol.* **2019**, *102*, 134–139. [CrossRef]
100. Yang, Z.; Shi, W.; Yang, W.; Liang, L.; Yao, W.; Chai, L.; Gao, S.; Liao, Q. Combination of bioleaching by gross bacterial biosurfactants and flocculation: A potential remediation for the heavy metal contaminated soils. *Chemosphere* **2018**, *206*, 83–91. [CrossRef] [PubMed]
101. Wen, Y.M.; Wang, Q.P.; Tang, C.; Chen, Z.L. Bioleaching of heavy metals from sewage sludge by *Acidithiobacillus thiooxidans*—A comparative study. *J. Soils Sediments* **2012**, *12*, 900–908. [CrossRef]
102. Ancona, V.; Caracciolo, B.; Campanale, C.; Rascio, I.; Grenni, P.; Di Lenola, M.; Bagnuolo, G.; Felice, V. Heavy metal phytoremediation of a poplar clone in a contaminated soil in southern Italy. *J. Chem. Biotechnol.* **2019**, *95*, 940–949. [CrossRef]
103. Zacchini, M.; Pietrini, F.; Mugnozza, G.S.; Iori, V.; Pietrosanti, L.; Massacci, A. Metal Tolerance, Accumulation and Translocation in Poplar and Willow Clones Treated with Cadmium in Hydroponics. *Water Air Soil Pollut.* **2009**, *197*, 23–34. [CrossRef]
104. Zacchini, M.; Iori, V.; Scarascia Mugnozza, G.; Pietrini, F.; Massacci, A. Cadmium accumulation and tolerance in Populus nigra and Salix alba. *Biol. Plant.* **2011**, *55*, 383–386. [CrossRef]
105. Courchesne, F.; Turmel, M.-C.; Cloutier-Hurteau, B.; Constantineau, S.; Munro, L.; Labrecque, M. Phytoextraction of soil trace elements by willow during a phytoremediation trial in Southern Québec, Canada. *Int. J. Phytoremediat.* **2017**, *19*, 545–554. [CrossRef]
106. Eskandari, H.; Amraie, A.A. Ability of some crops for phytoremediation of nickel and zinc heavy metals from contaminated soils. *J. Adv. Environ. Health Res.* **2016**, *4*, 234–239. [CrossRef]
107. Jia, W.; Lv, S.; Feng, J.; Li, J.; Li, Y.; Li, S. Morphophysiological characteristic analysis demonstrated the potential of sweet sorghum (*Sorghum bicolor* (L.) Moench) in the phytoremediation of cadmium-contaminated soils. *Environ. Sci. Pollut. Res.* **2016**, *23*, 18823–18831. [CrossRef]
108. Dinh, T.T.T. Evaluating the growth capacity and heavy metal absorption of sweet sorghum and grain sorghum at the seedling stage. *J. Agric. Dev.* **2018**, *17*, 44–48.
109. Abdulhamid, Y.; Abdulsalam, M.S.; Matazu, I.K.; Suleiman, A.B. Phytoremediation of some heavy metals in industrially contaminated dumpsite soil by hyptis suaveolens and physalis philadelphica. *Katsina J. Nat. Appl. Sci.* **2017**, *6*, 168–178.
110. Yang, Y.; Zhou, X.; Tie, B.; Peng, L.; Li, H.; Wang, K.; Zeng, Q. Comparison of three types of oil crop rotation systems for effective use and remediation of heavy metal contaminated agricultural soil. *Chemosphere* **2017**, *188*, 148–156. [CrossRef] [PubMed]
111. Saran, A.; Fernandez, L.; Cora, F.; Savio, M.; Thijs, S.; Vangronsveld, J.; Merini, L.J. Phytostabilization of Pb and Cd polluted soils using Helianthus petiolaris as pioneer aromatic plant species. *Int. J. Phytoremediat.* **2020**, *22*, 459–467. [CrossRef]
112. Mahmood-ul-Hassan, M.; Yousra, M.; Saman, L.; Ahmad, R. Floriculture: Alternate non-edible plants for phyto-remediation of heavy metal contaminated soils. *Int. J. Phytoremediat.* **2020**, *22*, 725–732. [CrossRef] [PubMed]
113. Chirakkara, R.A.; Reddy, K.R.; Cameselle, C. Electrokinetic Amendment in Phytoremediation of Mixed Contaminated Soil. *Electrochim. Acta* **2015**, *181*, 179–191. [CrossRef]

114. Agnello, A.C.; Bagard, M.; van Hullebusch, E.D.; Esposito, G.; Huguenot, D. Comparative bioremediation of heavy metals and petroleum hydrocarbons co-contaminated soil by natural attenuation, phytoremediation, bioaugmentation and bioaugmentation-assisted phytoremediation. *Sci. Total Environ.* **2016**, *563–564*, 693–703. [CrossRef]
115. Sigua, G.C.; Novak, J.M.; Watts, D.W.; Ippolito, J.A.; Ducey, T.F.; Johnson, M.G.; Spokas, K.A. Phytostabilization of Zn and Cd in Mine Soil Using Corn in Combination with Biochars and Manure-Based Compost. *Environments* **2019**, *6*, 69. [CrossRef]
116. Hussein, H.S.; Ruiz, O.N.; Terry, N.; Daniell, H. Phytoremediation of Mercury and Organomercurials in Chloroplast Transgenic Plants: Enhanced Root Uptake, Translocation to Shoots and Volatilization. *Environ. Sci. Technol.* **2007**, *41*, 8439–8446. [CrossRef] [PubMed]
117. Sakakibara, M.; Watanabe, A.; Sano, S.; Inoue, M.; Kaise, T. Phytoextraction and phytovolatilization of arsenic from As-contaminated soils by *Pteris vittata*. In Proceedings of the Annual International Conference on Soils, Sediments, Water and Energy, Amherst, MA, USA, 16–19 October 2006; Volume 12.
118. Dermont, G.; Bergeron, M.; Mercier, G.; Richer-Laflèche, M. Soil washing for metal removal: A review of physical/chemical technologies and field applications. *J. Hazard. Mater.* **2008**, *152*, 1–31. [CrossRef]
119. Dos Santos, J.V.; Varón-López, M.; Soares, C.R.F.S.; Lopes Leal, P.; Siqueira, J.O.; de Souza Moreira, F.M. Biological attributes of rehabilitated soils contaminated with heavy metals. *Environ. Sci. Pollut. Res.* **2016**, *23*, 6735–6748. [CrossRef] [PubMed]
120. Nejad, Z.D.; Jung, M.C.; Kim, K.-H. Remediation of soils contaminated with heavy metals with an emphasis on immobilization technology. *Environ. Geochem. Health* **2018**, *40*, 927–953. [CrossRef]
121. Seshadri, B.; Bolan, N.S.; Choppala, G.; Kunhikrishnan, A.; Sanderson, P.; Wang, H.; Currie, L.D.; Tsang, D.C.W.; Ok, Y.S.; Kim, G. Potential value of phosphate compounds in enhancing immobilization and reducing bioavailability of mixed heavy metal contaminants in shooting range soil. *Chemosphere* **2017**, *184*, 197–206. [CrossRef]
122. Zhang, G.; Guo, X.; Zhao, Z.; He, Q.; Wang, S.; Zhu, Y.; Yan, Y.; Liu, X.; Sun, K.; Zhao, Y.; et al. Effects of biochars on the availability of heavy metals to ryegrass in an alkaline contaminated soil. *Environ. Pollut.* **2016**, *218*, 513–522. [CrossRef] [PubMed]
123. Lu, K.; Yang, X.; Gielen, G.; Bolan, N.; Ok, Y.S.; Niazi, N.K.; Xu, S.; Yuan, G.; Chen, X.; Zhang, X.; et al. Effect of bamboo and rice straw biochars on the mobility and redistribution of heavy metals (Cd, Cu, Pb and Zn) in contaminated soil. *J. Environ. Manag.* **2017**, *186*, 285–292. [CrossRef]
124. Xing, J.; Li, L.; Li, G.; Xu, G. Feasibility of sludge-based biochar for soil remediation: Characteristics and safety performance of heavy metals influenced by pyrolysis temperatures. *Ecotoxicol. Environ. Saf.* **2019**, *180*, 457–465. [CrossRef] [PubMed]
125. Jeong, H.Y.; Klaue, B.; Blum, J.D.; Hayes, K.F. Sorption of Mercuric Ion by Synthetic Nanocrystalline Mackinawite (FeS). *Environ. Sci. Technol.* **2007**, *41*, 7699–7705. [CrossRef] [PubMed]
126. Gueye, M.T.; Di Palma, L.; Allahverdeyeva, G.; Bavasso, E.; Stoller, M.; Vilardi, G. The Influence of Heavy Metals and Organic Matter on Hexavalent Chromium Reduction by Nano Zero Valent Iron in Soil. *Chem. Eng. Trans.* **2019**, *47*, 289–294. [CrossRef]
127. Skyllberg, U.; Drott, A. Competition between Disordered Iron Sulfide and Natural Organic Matter Associated Thriols for Mercury(II)—An EXAFS Study. *Environ. Sci. Technol.* **2010**, *44*, 1254–1259. [CrossRef] [PubMed]
128. Wang, T.; Liu, Y.; Wang, J.; Wang, X.; Liu, B.; Wang, Y. *In-situ* remediation of hexavalent chromium contaminated groundwater and saturated soil using stabilized iron sulfide nanoparticles. *J. Environmental Management* **2019**, *231*, 679–686. [CrossRef]
129. Jin, L.N.; Qian, X.Y.; Wang, J.G.; Aslan, H.; Dong, M. MIL-68 (In) nano-rods for the removal of Congo red dye from aqueous solution. *J. Colloid Interface Sci.* **2015**, *453*, 270–275. [CrossRef]
130. Jin, K.; Lee, B.; Park, J. Metal-organic frameworks as a versatile platform for radionuclide management. *Coord. Chem. Rev.* **2021**, *427*, 213473. [CrossRef]
131. Zhang, W.; Tong, L.; Yuan, Y.; Liu, Z.; Huang, H.; Tan, F.; Qiu, R. Influence of soil washing with a chelator on subsequent chemical immobilization of heavy metals in a contaminated soil. *J. Hazard. Mater.* **2010**, *178*, 578–587. [CrossRef] [PubMed]
132. He, F.; Gao, J.; Pierce, E.; Strong, P.J. In situ remediation technologies for mercury-contaminated soil. *Environ. Sci. Pollut. Res.* **2015**, *22*, 8124–8147. [CrossRef] [PubMed]
133. Reddy, K.R.; Cameselle, C. *Electrochemical Remediation Technologies for Polluted Soils, Sediments and Groundwater*; John Wiley & Sons, Inc.: Hoboken, NJ, USA, 2009; ISBN 978-0-470-38343-8.
134. Sun, Z.; Wu, B.; Guo, P.; Wang, S.; Guo, S. Enhanced electrokinetic remediation and simulation of cadmium-contaminated soil by superimposed electric field. *Chemosphere* **2019**, *233*, 17–24. [CrossRef] [PubMed]
135. Zabochnicka-Świątek, M.; Krzywonos, M. Potentials of Biosorption and Bioaccumulation Processes for Heavy Metal Removal. *Pol. J. Environ. Stud.* **2014**, *23*, 551–561.
136. Velkova, Z.; Kirova, G.; Stoytcheva, M.; Kastadinova, S.; Todorova, K.; Gochev, V. Immobilized microbial biosorbents for heavy metals removal. *Eng. Life Sci.* **2018**, *18*, 871–881. [CrossRef] [PubMed]
137. Nwaichi, E.O.; Dhankher, O.P. Heavy Metals Contaminated Environments and the Road Map with Phytoremediation. *J. Environ. Prot.* **2016**, *7*, 41–51. [CrossRef]
138. Dumbrava, A.; Birghila, S.; Munteanu, M. Contributions on enhancing the copper uptake by using natural chelators, with applications in soil phytoremediation. *Int. J. Environ. Sci. Technol.* **2015**, *12*, 929–938. [CrossRef]
139. Shackira, A.M.; Puthur, J.T. Phytostabilization of Heavy Metals: Understanding of Principles and Practices. In *Plant-Metal Interactions*; Srivastava, S., Srivastava, A., Suprasanna, P., Eds.; Springer: Cham, Switzerland, 2016; online; ISBN 9783030207328.
140. Pérez, A.P.; Sánchez, S.P.; Van Liedekerke, M. *Remediation Sites and Brownfields. Success Stories in Europe*; EUR 27530 EN; Publications Office of the European Union: Luxembourg, 2015.

141. Kim, B.-K.; Park, G.-Y.; Jeon, E.-K.; Jung, J.-M.; Jung, H.-B.; Ko, S.-H.; Baek, K. Field Application of In Situ Electrokinetic Remediation for As-, Cu-, Pb-Contaminated Paddy Soil. *Water Air Soil Pollut.* **2013**, *224*, 1698. [CrossRef]
142. Geoengineer. Available online: https://www.geoengineer.org (accessed on 2 December 2020).
143. Epelde, L.; Burges, A.; Mijangos, I.; Garbisu, C. Microbial properties and attributes of ecological relevance for soil quality monitoring during a chemical stabilization field study. *Appl. Soil Ecol.* **2014**, *75*, 1–12. [CrossRef]
144. Guo, X.; Wei, Z.; Wu, Q.; Li, C.; Qian, T.; Zheng, W. Effect of soil washing with only chelators or combining with ferric chloride on soil heavy metal removal and phytoavailability: Field experiments. *Chemosphere* **2016**, *147*, 412–419. [CrossRef]
145. Pietrosanti, L.; Pietrini, F.; Zacchini, M.; Matteucci, G.; Massacci, A.; Nardella, A.; Capotorti, G. Phytoremediation of a metal contaminated industrial soil of Porto Marghera by a short rotation forestry stand. In Proceedings of the 11th International Conference on Environmental Science and Technology Chania, Crete, Greece, 3–5 September 2009; pp. B-738–B-744.
146. Chen, W.; Li, H. Cost-Effectiveness Analysis for Soil Heavy Metal Contamination Treatments. *Water Air Soil Pollut.* **2018**, *229*. [CrossRef]

Article

Different Approaches for Incorporating Bioaccessibility of Inorganics in Human Health Risk Assessment of Contaminated Soils

Daniela Zingaretti * and Renato Baciocchi

Laboratory of Environmental Engineering, Department of Civil Engineering and Computer Science Engineering, University of Rome "Tor Vergata", Via del Politecnico 1, 00133 Rome, Italy; baciocchi@ing.uniroma2.it
* Correspondence: zingaretti@ing.uniroma2.it

Abstract: Ingestion of soil represents one of the critical exposure pathways in the human health risk assessment (HHRA) framework at sites contaminated by inorganic species, especially for residential scenarios. HHRA is typically carried out through starting from the so-called "total concentration", which is estimated from the fraction of inorganic species extracted from the soil using standardized approaches, i.e., microwave acid extraction. Due to the milder conditions, a smaller portion of the inorganics present in the soil is actually dissolved in the gastro-intestinal tract (bioaccessible fraction), and afterward reaches the bloodstream, exerting an effect on human health (bioavailable fraction). Including bioaccessibility in HHRA could then allow for the achievement of a more realistic assessment than using the total concentration. In this paper, the bioaccessible concentration of different inorganics in soil samples collected from a firing range was estimated by applying two in vitro tests, i.e., the Unified Barge Method (UBM) and the Simple Bioaccessibility Extraction Test (SBET). Moreover, different options for incorporating bioaccessibility in HHRA for the estimation of the cleanup goals were also applied and discussed. Despite the notable differences in terms of reagents and procedure between the two methods, the obtained results were quite close, with the SBET method providing slightly higher values. The role of the soil particle size distribution on the calculation of the cleanup goals accounting for bioaccessibility is also discussed.

Keywords: bioaccessibility; bioavailability; metals; contaminated sites; risk assessment

1. Introduction

Incidental ingestion of contaminated soils represents one of the main exposure pathways when assessing human health risks associated with metals or metalloids such as Pb, As, and Cd [1]. This is particularly true at sites characterized by residential or recreational use, due to the presence of children (e.g., [2]). In fact, for their typical hand-to-mouth behavior, the fine fraction of soil could adhere to children's hand and then be ingested [3,4]. Once ingested, the contaminants present in the soil can be mobilized by the gastrointestinal juices (bioaccessibile fraction) and then adsorbed, i.e., pass the intestinal wall reaching the blood circulation (bioavailable fraction) [5,6].

According to the ASTM-RBCA (American Society for Testing and Materials -Risk-Based Corrective Action) approach [7,8], human health risk assessment (HHRA) for the soil ingestion pathway is typically carried out starting from the total soil concentration of the contaminants, which is estimated from the fraction of inorganic species extracted from the soil using standardized approaches, i.e., microwave acid extraction. Combining it with the relevant exposure factors, i.e., the ingestion rate (IR), the exposure duration (ED), and the body weight (BW), allows for the estimation of the exposure (E) for a given receptor.

There is evidence in the literature that this assumption can lead to overestimating the actual risks for human health [9]. Indeed, only a fraction of the inorganic content of a soil has been shown to reach the blood circulation system once ingested [3]. Furthermore, it is worth noting that the bioavailable fraction of a contaminant is strictly correlated to the soil properties and contaminant characteristics, which change from site to site [10–12].

Different methods, either in vivo or in vitro, have been successfully developed and tested for assessing the bioavailability of metals and metalloids in soils. The in vivo procedures use swine and mouse as models for evaluating the inorganic accumulation in animal tissues or blood after exposure to contaminated soils and have been proven to be quite robust (e.g., [13,14]). However, the adoption of these methodologies within any HHRA procedure is impractical due to the high costs and times required and to ethical considerations [2]. Several simple, cheap, and reproducible in vitro tests have also been developed, which allow for the estimation of the bioaccessible fraction of a contaminant, and consequently provide a conservative estimate of the bioavailable fraction [4]. The bioavailable fraction will then be at most equal to the bioaccessible one. The different in-vitro tests rely on various reagents for mimicking at laboratory scale the digestive juices and conditions of the gastrointestinal tract of the human body so as to investigate the mobilization of the contaminants from the soil [15]. The methods investigated in the literature include the Physiologically Based Extraction Test (PBET) [10], the Simple Bioaccessibility Extraction Test (SBET) [16,17], the In Vitro Gastrointestinal (IVG) test [18], the methods developed by the Deutsches Institut für Normung e.V. [19] and by the Dutch Institute of Public Health (RIVM) [20], and finally the Unified Barge Method (UBM) [6]. Many differences can be found in comparing the operating conditions adopted by these methods. For instance, several methods (e.g., RIVM and DIN) are carried out without mixing, whereas others (e.g., UBM and SBET) are carried out using dynamic conditions. Furthermore, some methods (e.g., SBET) mimic only the gastric phase, which is assumed to be the most relevant one for the inorganic mobilization, whereas other procedures (e.g., UBM) try to simulate the mouth, the stomach, and the intestinal tract. All the above-mentioned tests have shown good in vivo–in vitro correlations (IVIVC) for the tested inorganics (e.g., [18,21,22]). However, the results obtained applying each method to the same soil can be significantly different in terms of the estimated bioaccessible fractions due to the diverse experimental conditions adopted by each procedure [5].

Two of the most used methods are the UBM and SBET, which present huge differences in terms of extraction juices, time, and complexity. The UBM method was developed by the Bioaccessibility Research Group of Europe (BARGE), which was looking for a joint decision on a harmonized bioaccessibility method. Hence, it was chosen to adopt the procedure previously elaborated by RIVM. The UBM procedure was tested in an inter-laboratory trial [23]) and was validated in vivo for arsenic, cadmium, and lead. The UBM procedure is physiologically based and is aimed at mimicking mouth, stomach, and small intestinal cavities. Therefore, this method includes the preparation of four digestive fluids (i.e., saliva, gastric, duodenal, and simulated bile solutions) using both chemical reagents and enzymes that are sequentially put in contact at 37 °C with a soil sample in two extraction phases. Although this procedure effectively simulates at the laboratory scale the process that the soil undergoes once ingested, the preparation of the digestive solution is complex and the procedure can be labor intensive.

The SBET represents a simplified form of the PBET procedure developed in response to some requests from USEPA (United States Environmental Protection Agency) regional offices [15]. Unlike the UBM method, the SBET is not physiologically based and has no digestive enzyme in the gastric fluid. Namely, the procedure consists in a single extraction step with a 0.4 M glycine (pH = 1.5) solution at 37 °C for 1 h simulating the stomach acidic environment. This method was initially validated against swine in vivo assay for lead [24,25]. However, afterwards the SBET procedure was also applied for measuring the bioaccessibile fraction of cadmium and arsenic [5] and more recently was also applied to other metals or metalloids (e.g., [26,27]). Currently it is adopted by the USEPA as a standard procedure for evaluating the bioaccessibility of lead and arsenic [17].

The availability of standard in vitro procedures allowed for the incorporation of oral bioaccessibility into forward HHRA on soils [28] and toxic waste [29] for estimating risks for exposed receptors and into backward HHRA for the definition of soil cleanup goals [30]. However, the final outcome of the HHRA procedure depends on the bioaccessibility method used but also on how bioaccessibility is incorporated into HHRA. First, one aspect to consider is that despite the characterization of soil contamination usually being carried out on the particle size fraction below 2 mm, the different in vitro procedures are typically performed on finer size fractions, i.e., the <250 µm fraction for the UBM and the <150 µm for the SBET. Hence, it is crucial to decide to which total metal content to refer the bioaccessible concentration. According to Guney et al. [28], considering the total inorganic content of the fine fraction would provide a more appropriate and conservative estimate of the concentration in ingested soil since the fine fraction tends to have higher inorganic concentrations in general and is more likely to be ingested by children. More recently, Mehta et al. [29] also performed both the bioaccessibility and the risk calculations on the fine fraction, for consistency, as they carried out the bioaccessibility tests on this fraction using the UBM. Bioaccessibility was found to depend on the soil particle size considered [31], but Li et al. [32] concluded that although a general trend of higher potentially toxic element (PTE) bioaccessibility in finer fraction is found, this does not mean that the highest PTE bioaccessibility is always found in the finest size fraction. How to incorporate the bioaccessible fraction estimated by the standard in vitro procedure into forward and backward HHRA is still a matter of debate, especially with reference to how to account for the distribution of the contaminants and for the trend of bioaccessibility among the different soil particle size fractions.

Shooting ranges have long been recognized as a potential source for environmental contamination due to the large accumulation of lead (Pb) in soil (e.g., [33–38]) as a result of firing activities with bullets that are mainly composed of lead [39]. There are many thousands of shooting ranges operating around the world for recreational activity and military training, which are also characterized by the presence of co-contaminants, such as antimony, zinc, copper, and nickel, [40]. Since soil ingestion represents a critical exposure route, especially for adults and children using the site for recreation, bioaccessibility may allow for the more correct estimation of the cleanup goals, also considering that lead is one of the elements on which the in vitro bioaccessibility methods discussed above have been validated.

In this work, we investigated the bioaccessibility of metals and metalloids in soil samples collected at a firing range through applying the two procedures discussed above, i.e., the UBM and the SBET. Lead but also other metals and metalloids present in the site were considered as target contaminants. The estimated bioaccessibilities were then incorporated into a HHRA procedure carried out in backward mode, thus allowing us to estimate the cleanup goals. The results obtained using different methods for estimating and incorporating bioaccessibility into HHRA allowed us to assess how different choices may affect the outcome of the overall risk assessment.

2. Materials and Methods

2.1. Soil Characterization

The soil samples used in this work were collected at a military firing range in central Italy, which is still in operation for about 4 months in a year. The definition of the cleanup goals and of the cleanup approach is underway. Currently, a part of the site is open to the general public for leisure activity during the weekends and holidays. Four samples of topsoil, named A1, A2, S1, and S2, were collected at a depth of 0–0.3 m within an investigation campaign performed at the site. These samples presented high concentrations of some metals, mainly lead, which, as is well known, is related to the firing activity performed at this type of site (e.g., [33–36]). The soil samples were dried (40 °C) for at least 4 days and weighted. Afterwards, the samples were sieved with stainless steel sieves to 2 mm, 250 µm, and 150 µm so as to obtain the soil fractions suitable for characterization analysis and bioaccessibility tests. All 3 fractions were analyzed for the total inorganic content, according to USEPA 3051A method [41]. Namely, the soil samples underwent a microwave-assisted acid digestion with nitric acid (HNO$_3$) and hydrochloric acid (HCl), and the obtained solutions were analyzed by inductively coupled plasma optical emission spectroscopy (ICP-OES; Varian 710-ES) for measuring As, Be, Cd, Cr, Ni, Pb, V, and Zn concentrations. The fractions with particle size below 250 and 150 μm were used for measuring the bioaccessible concentration using the UBM and SBET, respectively, as described in Section 2.2. All analysis were performed in triplicate.

2.2. Bioaccessible Concentration

The procedure adopted for carrying out the UBM tests was the one reported in the International Standard ISO 17924 [6]. The soil fraction used in the UBM method was the one with a diameter lower than 250 μm, since this portion of soil is believed to adhere to human hands and become ingested during a hand-to-mouth activity [3].

The 4 extraction fluids, i.e., saliva, gastric, duodenal, and bile solutions were prepared the day before performing the tests according to the specific composition and indications reported in the above-mentioned standard. Each one of these fluids was prepared by mixing a 500 mL solution with inorganics to a 500 mL solution with organic and solid constituents, with a composition reported in detail in the ISO 17924 Standard [6]. Briefly, the UBM includes 2 sequential extraction steps, corresponding to the gastric and gastrointestinal phases. In the first step, the gastric and gastrointestinal samples were prepared by adding 0.6 g of soil to 9 mL of saliva and 13.5 mL of gastric fluid; then, after checking that the pH of the solution was within the range 1.2 ± 0.05, the samples were incubated in an end-over-end shaker at 37 °C for 1 h. At the end of this stage, we verified that the pH of each suspension (both gastric and gastrointestinal samples) was lower than 1.5. Afterwards, the gastric samples were centrifuged at 4500× g for 15 min, and the supernatants were collected for analysis by ICP-OES. The gastrointestinal samples were mixed with 27 mL of simulated duodenal fluid and 9 mL of simulated bile fluid, and after checking the pH to be in the range 6.3 ± 0.5, we incubated them in an end-over-end rotator at 37 °C for 4 h. At the end of this phase, after controlling the pH value (that should be 6.3 ± 0.5), we centrifuged the suspensions at 4500× g for 15 min and analyzed the solutions by ICP-OES.

The bioaccessibility tests performed adopting the SBET were carried out in agreement with the USEPA standard operating procedure [17]. It is worth noting that, differently from the UBM method, following the recommendation for the assessment of incidental ingestion of soil developed by USEPA [42], the soil used in the SBET method was the fraction with diameter lower than 150 μm, which, on the basis of studies published in the last years, represent the dominant fraction for dermally adhered soil (e.g., [43]). The extraction fluid was represented by a 0.4 M glycine solution whose pH was adjusted to 1.5 by adding HCl. Then, 1 g of the soil sample was put in contact with 100 mL of the extraction fluid and rotated end-over-end at 37 °C at 30 rpm for 1 h. At the end of the experiment, we checked that the final pH of the solution was in the range 1.5 ± 0.5. Afterwards, the soil–liquid solution was allowed to settle the solid particles, and then the liquid was filtered through a 0.45 μm cellulose acetate disk before analysis by ICP-OES.

All reagents used for performing characterization analysis and bioaccessibility tests were of analytical grade and were purchased by Sigma-Aldrich (St. Louis, MO, USA).

2.3. Quality Control

In order to assure the correct application of the complex procedure envisioned in the UBM, prior to performing the tests on the soil samples collected in the firing range, we applied the in vitro procedure on a certified reference material, i.e., the BGS102 provided by the British Geological Survey (BGS). This reference material is ferritic brown earth collected from North Lincolnshire that was ball-milled to a particle size lower than 40 μm. The main chemical and physical properties of this reference material, retrieved from the Certificate of Analysis provided by BGS, are reported in Table S1 in the Supplementary Information. For comparison purposes, the SBET was also performed on the same material.

For quality control, the UBM or SBET extractions on the soil samples A1, A2, S1, and S2 collected in the firing range were carried out in triplicate, and for each batch of extractions, 3 procedural blanks were also tested. The detection limits of the ICP-OES analysis were 0.033 mg/L for As and 0.005 mg/L for the other analyzed elements. Hence, the limits of quantification for the total content determination were equal to 1.58 mg/kg for As and 0.24 mg/kg for the other elements. As far as the bioaccessibility tests are concerned, considering the amount of soil and liquid used in each procedure, the limit of quantification was estimated to be equal to 1.24 mg/kg for As, 0.19 mg/kg for the other elements in the gastric phase of the UBM, and 3.3 mg/kg for As and 0.5 mg/kg for the other elements in the gastrointestinal phase of the UBM and in the SBET.

2.4. Risk Assessment

Incorporating bioaccessibility requires the modification of the approach usually undertaken to carry out HHRA at contaminated sites. In this discussion, we make reference to the approach followed in Italy [44], which essentially relies on the RBCA-ASTM approach [7,8]. Namely, in forward mode, Risk or Hazard Index is estimated from the total soil concentration (C), which is calculated by normalizing the concentration measured on the soil particle size fraction below 2 mm ($C_{TOT\ 2mm}$) to the soil fraction with particle diameter between 2 mm and 2 cm (F_{2mm}):

$$C = C_{TOT2mm} \cdot F_{2mm} \tag{1}$$

This means to assume that all contaminants are present in the particle size fraction below 2 mm, whereas the fraction between 2 mm and 2 cm is assumed to be clean.

For the soil ingestion pathway, risk (R) and hazard index (HI) are given by

$$R = EM \cdot SF \cdot C \tag{2}$$

$$HI = \frac{EM \cdot C}{RfD} \tag{3}$$

where SF $(mg/kg/d)^{-1}$ and RfD (mg/kg/d) are the carcinogenic slope factor and reference dose of a given chemical, respectively.

EM, which includes the exposure parameters, is given by

$$EM = \frac{ED \cdot EF \cdot IR}{BW \cdot AT} \tag{4}$$

where ED is the exposure duration (y), EF the exposure frequency (d/y), IR the soil ingestion rate (mg/d), BW the body weight (kg), and AT the averaging time (y).

The backward HHRA allows for the estimation of the cleanup goal for a given chemical, here named CSR by the Italian legislation [45], by simply reversing the risk equations shown above, thus leading to

$$CSR_{canc} = \frac{TR}{EM \cdot SF} \tag{5}$$

$$CSR_{non\ canc} = \frac{THI \cdot RfD}{EM} \tag{6}$$

where TR and THI are the target carcinogenic risk (10^{-6} for the Italian legislation) and the target hazard index (1 for the Italian legislation), respectively. When incorporating the bioaccessibility (BA) in the estimate of the cleanup goals for the soil ingestion pathway, we obtained the following equations:

$$CSR_{canc} = \frac{TR}{EM \cdot SF \cdot BA} \tag{7}$$

$$CSR_{non\ canc} = \frac{THI \cdot RfD}{EM \cdot BA} \tag{8}$$

The question now becomes how to estimate the bioaccessible fraction of hazardous compounds to be considered in the above-mentioned equations.

One option proposed in the literature (e.g., [28,29]) relies on estimating the bioaccessibility with reference to the total concentration measured in the same particle size fraction used for measuring the bioaccessible concentration, as shown in the following equation.

$$BA_1 = \frac{C_{BA}}{C_{TOT\,FINE}} \cdot 100 \qquad (9)$$

where C_{BA} (mg kg^{-1}) is the bioaccessible concentration of each inorganic measured adopting for instance the UBM or SBET methods, and $C_{TOT\,FINE}$ is the total inorganic content (mg kg^{-1}) measured on the same soil fraction used for the bioaccessibility test (i.e., d < 250 μm for the UBM or d < 150 μm for the SBET).

Another option could be to calculate the bioaccessibility (BA) with reference to the total soil concentration (C), i.e., the concentration in the soil fraction < 2 mm normalized to the to the soil fraction with particle diameter between 2 mm and 2 cm, which is usually considered in the HHRA procedure. However, for consistency, in this case, the particle size distribution also had to be taken into account, normalizing the bioaccessible concentration to the soil fraction used in the in vitro methods, thus leading to the following equation:

$$BA_2 = \frac{C_{BA} \cdot F_{BA}}{C} \cdot 100 \qquad (10)$$

where C_{BA} (mg kg^{-1}) is again the bioaccessible concentration of each element measured adopting either the UBM or the SBET method, F_{BA} is the soil fraction used in the in vitro methods (i.e., d < 250 μm for the UBM or d < 150 μm for the SBET), and C is the total soil concentration (Equation (1)).

3. Results and Discussion

3.1. Soil Characterization

Table 1 shows the particle size distribution of the different soil samples, with reference to the 2 mm, 250 μm, and 150 μm sieves.

Table 1. Particle size fractions of the soil samples collected in the firing range.

	A1	A2	S1	S2
	(%)	(%)	(%)	(%)
d > 2 mm	87.8	31.2	57.7	60.7
d < 2 mm	12.2	68.6	42.3	39.3
d < 250 μm	2.4	11	4.2	3.8
d < 150 μm	n.d.	4.3	2.2	2.1

Sample A1 is mostly made of coarse particles, above 2 mm, which correspond to 88% weight of the overall sample, whereas the fraction below 250 μm represents only 2.4% weight. In this case, the particle size fraction below 150 μm was so low that it was not possible to collect an amount sufficient for carrying out the characterization and the bioaccessibility tests. Sample A2 is slightly finer, with 69% weight of the particles below 2 mm, whereas the particle size fractions below 250 μm and 150 μm were slightly higher than in A1, i.e., 11% and 4.3%, respectively. Samples S1 and S2, instead, had similar particle size distribution, with around 60% of the particles above 2 mm and the finer fractions representing around 4% (d < 250 μm) and 2% (d < 150 μm).

Figure 1 reports for each soil sample collected in the firing range (i.e., A1, A2, S1, and S2) the metal and metalloid contents determined applying USEPA Method 3051A on the fractions with particle diameters lower than 2 mm, 250 μm, and 150 μm. The contents shown in Figure 1 are the mean values obtained from each triplicate. As mentioned above, for sample A1, the amount of the soil fraction d < 150 µm was insufficient for performing the analysis.

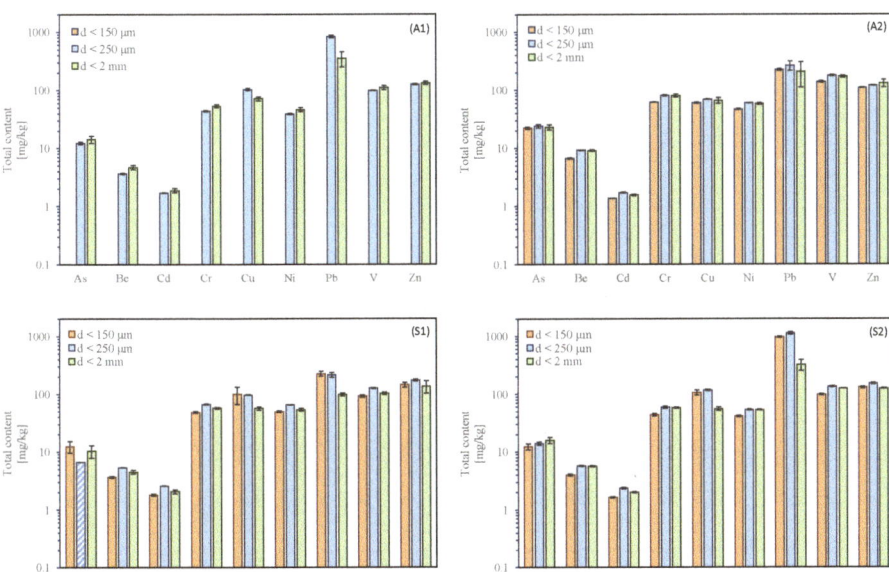

Figure 1. Total inorganic content of fraction d < 2 mm, d < 250 µm, and d < 150 µm for samples **A1, A2, S1,** and **S2** (striped bars represent values below the limit of quantification).

There were no substantial differences found between the concentrations measured in the three different soil fractions. Lead was the only inorganic that regularly presented a higher concentration in the finest fractions d < 250 µm and d < 150 µm than in fraction d < 2 mm. Indeed, the finer soil particles usually present a greater specific surface than the coarser fraction, and hence may sorb a higher amount of inorganic per unit weight [46]. Since this effect was observed on lead only, the enrichment of Pb observed in the finer fraction of the soil samples was linked to the firing activity carried out in the shooting range and to the bullet residues that can end up in soils after use [35]. This result is in agreement with what has also been observed in previous studies that analyzed the Pb contamination in firing ranges (e.g., [36]).

3.2. Quality Control

Figure 2 reports the bioaccessible concentration obtained adopting UBM and SBET procedures to the reference material BGS102. Furthermore, for quality control purposes, in the same figure, the maximum and minimum bioaccessible concentration of As and Pb reported in the Certificate of Analysis of this material adopting UBM are also reported. It is possible to observe that the values obtained in this work adopting the UBM on the d < 250 µm fraction fall within the range reported for As and Pb in the BGS certificate of analysis. This result confirms that all the different steps of the UBM procedure were correctly carried out.

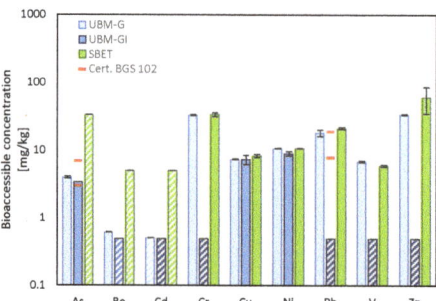

Figure 2. Bioaccessibile concentrations obtained applying the Unified Barge Method (UBM) (UBM-G: gastric phase; UBM-GI: gastro-intestinal phase) and Simple Bioaccessibility Extraction Test (SBET) to BGS102. The striped bars represent values below the limit of quantification. A 10-fold dilution of the extract for the SBET was applied.

3.3. Bioaccessible Concentration

Figure 3 reports the results of the UBM and SBET methods for the four samples collected at the firing range in terms of bioaccessible concentrations. Namely, for each sample, the concentrations measured in the gastric and gastrointestinal phases of the UBM (named UBM-G and UBM-GI, respectively) were compared with the values obtained by the SBET. For sample BGS102, the values of the bioaccessible concentration of As and Pb reported in the certificate of analysis provided by the supplier are also shown for comparison.

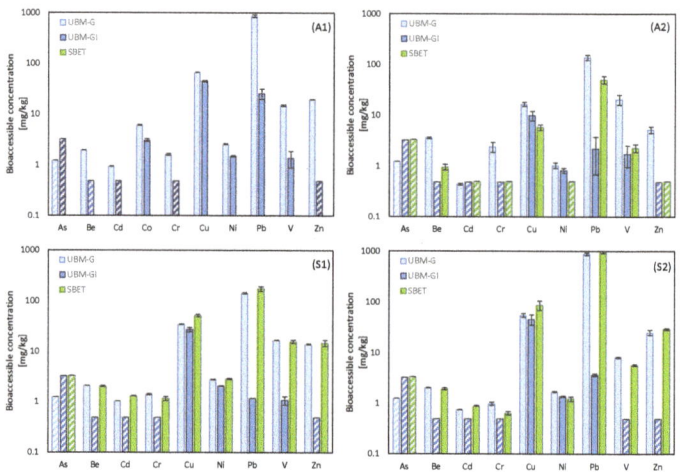

Figure 3. Comparison of the bioaccessibile concentrations obtained applying UBM (UBM-G: gastric phase; UBM-GI: gastro-intestinal phase) and SBET to samples **A1**, **A2**, **S1**, and **S2**. The striped bars represent values below the limit of quantification.

Regarding the UBM results, it is possible to observe that for all the tested samples, the bioaccessible concentration measured for the different inorganics in the gastric phase was higher than that obtained for the gastro-intestinal phase. This result is in agreement with what has already been observed in previously published papers (e.g., [2,15,46]) that attributed this behavior to the strong influence of the pH of the solution. The pH of the gastric phase in fact was significantly lower (pH = 1.2) compared to that of the gastrointestinal phase (pH = 6.3), leading to a higher mobilization of the inorganics from the soil to the solution. As also stated by Medlin et al. [16], the change in pH from low to neutral values passing from gastric to gastrointestinal conditions is likely to determine a precipitation of inorganics and hence a lower bioaccessibility.

Comparing the bioaccessible concentrations obtained for the two methods, one can observe that, except for A2, which was probably affected by heterogeneity, the SBET proved to be more conservative, with values slightly higher than those of the UBM. Furthermore Figure 3 shows that for all the analyzed inorganics, the bioaccessible concentrations determined by the SBET were quite close to the values obtained for the gastric phase of the UBM (i.e., UBM-G). These results could have been expected considering that, although the two methods are characterized by very different procedures and reagents, the pH of the solution in the SBET and in the gastric phase of UBM were similar (1.2 compared to 1.5). This observation underlines once more that pH is a key parameter to consider for assessing bioaccessibility of inorganics.

3.4. Implication for the Contaminated Site Management

As described in Section 2.4, the bioaccessibility was estimated for each soil sample starting from the bioaccessible concentrations measured with the two in vitro methods (i.e., UBM and SBET) and using the two different equations, (9) and (10). Figure 4 reports the comparison of the values of the bioaccessible fraction (BA$_1$-UBM, BA$_1$-SBET, BA$_2$-UBM, and BA$_2$-SBET) obtained for each soil sample. In the UBM case, the bioaccessible fraction was estimated by using the higher concentration among those measured in the gastric and the gastrointestinal phases, thus adopting a conservative approach.

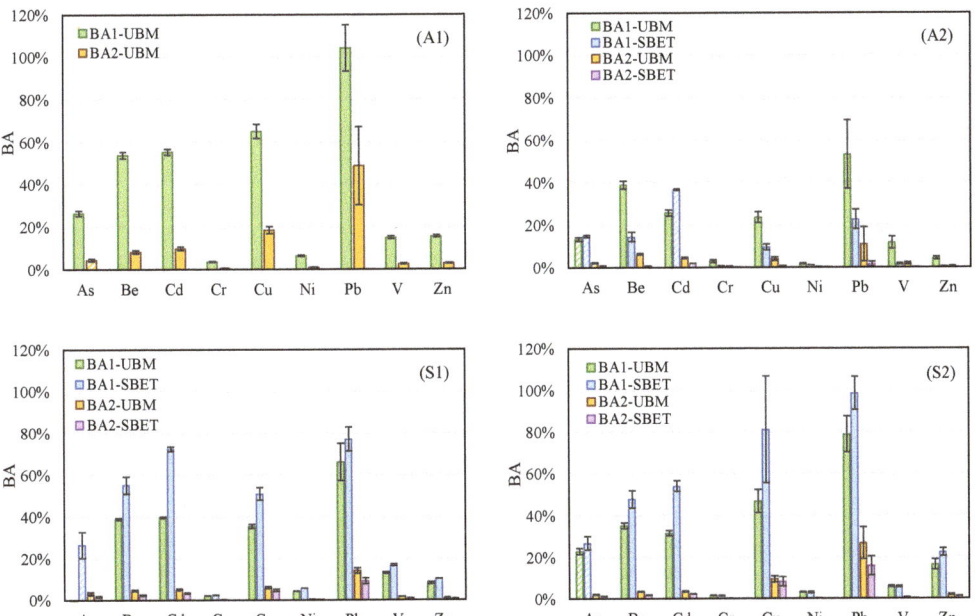

Figure 4. Bioaccessibility of metals and metalloids contained in samples **A1, A2, S1,** and **S2** estimated adopting Equations (9) and (10). Striped bars represent values obtained from bioaccessible concentration lower than the limit of quantification.

It has to be noticed that for the tested soil samples, the use of either the UBM or the SBET bioaccessible concentration did not notably affect the bioaccessibility when the results obtained with the same equation were compared. This means that, at least for the samples investigated in this work, despite the two methods following different procedures and reagents and being carried out on different size fractions, the estimated bioaccessibility is basically the same.

On the other hand, significantly different results were obtained using the two equations. Namely, the values obtained with Equation (10) (BA$_2$), regardless of the in vitro method used, were at least one order of magnitude lower than those calculated with Equation (9) (BA$_1$). This difference stems from the different assumptions behind the two equations. BA$_1$ did not account for the particle size distribution and was simply the ratio between the bioaccessible and total concentration measured on the same particle size fraction. In BA$_2$, the bioavailable concentration was normalized to the particle size fraction above 250 µm or 150 µm, for UBM and SBET, respectively, whereas the total concentration was also normalized to the soil fraction above 2 mm. As an example, we can look at the values obtained for lead in the different samples using the UBM. For sample A1, BA$_1$ was close to 1 and BA$_2$ around 0.5. This resulted from two counteracting effects (see Equation (10)). On the one hand, the particle size fraction below 250 µm (2.4%) was lower than the one below 2 mm (12.2%), i.e., a ratio around 0.2; on the other hand, the total lead concentration in the <250 µm fraction was higher than the one in the <2 mm fraction (836 mg kg^{-1} vs. 351 mg kg^{-1} respectively), i.e., a ratio of around 2.4. For sample A2, BA$_1$ was around 0.5, whereas BA$_2$ was around 0.1. In this case, the particle size fraction below 250 µm (11%) was lower than the one below 2 mm (68.6%), i.e., around 0.16 ratio, whereas the total lead concentration in the <250 µm fraction was slightly higher than the one in the <2 mm fraction (263 mg kg^{-1} vs. 206 mg kg^{-1}, respectively), i.e., a ratio of around 1.3. The behavior observed for samples S1 and S2 was somehow in between those observed for A1 and A2.

The bioaccessibility of lead in shooting ranges was investigated by numerous studies adopting different methods (see Table 2). Walraven et al. [35] investigated the Pb bioaccessibility in different sites characterized by various lead sources (Pb bullets and pellets, car battery Pb, gasoline Pb, diffuse Pb, made ground and city waste), observing that the shooting ranges showed the highest bioaccessibility. Generally, the bioaccessibility measured in shooting ranges falls within the range 45–100% [35–38]. The results obtained in the present study, particularly those estimated with Equation (9), fit well with data available in the literature. However, the values determined considering also the particle size distribution (hence using Equation (10)) were found to be generally lower than the values of previously published works. This difference can be explained through considering the fact that in these studies, the bioaccessibile fraction was estimated by referring to the total content of the soil.

Table 2. Studies on the bioaccessibility of Pb in shooting ranges.

	Bioaccessibility (%)	Method	References
Walraven et al., 2015	61–79	RIVM	[35]
Sanderson et al., 2012	46–70	SBET	[36]
Smith et al., 2011	50–100	SBRC	[37]
Moseley et al., 2008	75–85	PBET	[38]

The cleanup goals (CSR) for each soil sample were calculated using Equations (7) and (8), accounting for the bioaccessibility estimated using the two proposed approaches, i.e., either by Equation (9) (BA$_1$) or Equation (10) (BA$_2$). The obtained values are reported in Figure 5 for each soil sample and compared with the cleanup goal of each contaminant assuming 100% bioaccessibility. The latter value was obviously the same for all samples, since it depended only on the target risk (THI = 1 or TR = 10^{-6}), on the exposure parameters, and on the toxicity of the contaminant. For all contaminants and for any BA considered, the CSR were calculated for both the carcinogenic and non-carcinogenic effects, and the lower value was taken as representative CSR for each sample.

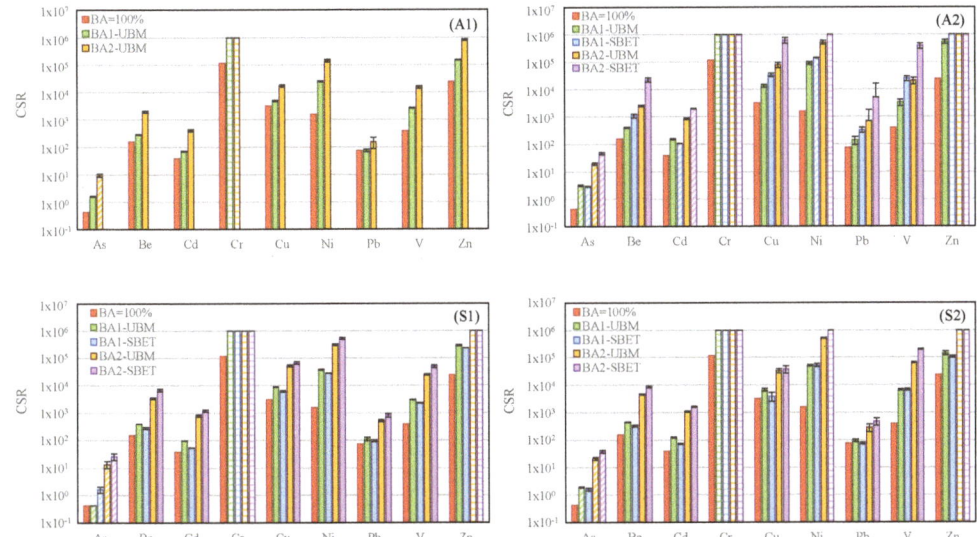

Figure 5. Comparison of the cleanup goals (CSR) calculated for each soil sample (**A1**, **A2**, **S1**, and **S2**) using the different bioaccessibility BA_1 and BA_2 or considering a 100% bioaccessibility. The striped bars indicate that the normalized cleanup goals were at least higher than the reported value, whereas the bars with horizontal strips indicate that the cleanup goal was higher than 10^6.

Clearly, the cleanup goals had an opposite behavior with respect to the one reported for bioaccessibility, meaning that a lower bioaccessibility corresponded to a higher cleanup goal. Let us consider the case of arsenic for sample A2—looking at Figure 5, we can observe that the CSR was higher when the estimated bioaccessibility, shown in Figure 5, was lower. In this case, BA was calculated by starting from a bioaccessible concentration equal to the quantification limit, and therefore represents a maximum estimate, i.e., the effective BA could be lower. Hence, the corresponding CSR represents a minimum estimate, and the effective CSR could be higher.

Similarly to what was observed for the bioaccessibility values, the cleanup goals proved to be significantly influenced by the approach adopted for evaluating the bioaccessible fraction. The bioaccessibility method did not notably affect the results, although the cleanup goals estimated adopting the SBET proved to be in most cases higher than those estimated on the basis of the UBM. Clearly, as expected, accounting for bioaccessibility led to a remarkable increase of the cleanup goals, up to 100-fold using BA_1 and up to 1000-fold using BA_2.

4. Conclusions

In this paper, the results obtained by adopting two different methods for estimating the bioaccessible concentration of inorganics were compared and discussed. Namely, four soil samples collected in a firing range were used to carry out tests adopting the UBM and SBET. The obtained results showed that among the extraction parameters, pH is one of the most important factors. Despite the huge differences in reagents, time, costs, and complexity of the two procedures adopted, the bioaccessibility using the SBET was only slightly higher than the one obtained with the gastric phase extraction of the UBM, which is characterized by similar pH values of the SBET. Hence, the latter method could be adopted at an early stage in order to obtain an easier and faster estimate of the bioaccessible concentration and the UBM could be applied in a second stage only on few samples that would benefit from a more accurate bioaccessibility estimate.

The bioaccessible concentrations estimated in the in vitro methods were then used for calculating the bioaccessible fractions and the cleanup goals, adopting two approaches, neglecting or considering the normalization on the basis of the particle size distribution, respectively. The first approach is surely more conservative but has the drawback that the cleanup goals related to the soil ingestion pathway would not depend on the soil particle size distribution; in principle, two soils with the same bioaccessible concentration but with different PSD (e.g., with 99% and 1% fine particles below 250 um, respectively) would have the same cleanup goal. The second approach allows us to overcome this drawback, but on the other hand could lead to excessively permissive cleanup goals.

How bioaccessibility should be incorporated into the HHRA framework needs further investigation on the source-to-hand and hand-to-mouth transport pathways, so as to avoid the use of either over- or under-protective risk assessment approaches.

Supplementary Materials: The following are available online at https://www.mdpi.com/article/10.3390/app11073005/s1, Table S1: Guidance values and confidence intervals reported in the certificate of Analysis of BGS102.

Author Contributions: Conceptualization, D.Z. and R.B.; methodology, D.Z. and R.B.; formal analysis, D.Z.; investigation, D.Z.; data curation, D.Z. and R.B.; writing—original draft preparation, D.Z. and R.B.; writing—review and editing, D.Z. and R.B.; visualization, D.Z.; supervision, R.B. All authors have read and agreed to the published version of the manuscript.

Funding: This research received no external funding.

Institutional Review Board Statement: Not applicable.

Informed Consent Statement: Not applicable.

Data Availability Statement: Not applicable.

Conflicts of Interest: The authors declare no conflict of interest.

References

1. Swartjes, F.A. Risk-Based Assessment of Soil and Groundwater Quality in the Netherlands: Standards and Remediation Urgency. *Risk Anal.* **1999**, *19*, 1235–1249. [CrossRef] [PubMed]
2. Li, H.B.; Li, M.Y.; Zhao, D.; Li, J.; Li, S.M.; Xiang, P.; Juhasz, A.L.; Ma, L.Q. Arsenic, lead, and cadmium bioaccessibility in contaminated soils: Measurements and validations. *Crit. Rev. Environ. Sci. Technol.* **2020**, *50*, 1303–1338. [CrossRef]
3. Ruby, M.V.; Davis, A.; Schoof, R.; Eberle, S.; Sellstone, C.M. Estimation of lead and arsenic bioavailability using a physiologically based extraction test. *Environ. Sci. Technol.* **1996**, *30*, 422–430. [CrossRef]
4. Denys, S.; Tack, K.; Caboche, J.; Delalain, P. Bioaccessibility, solid phase distribution, and speciation of Sb in soil sand in digestive fluids. *Chemosphere* **2008**, *74*, 711–716. [CrossRef]
5. Oomen, A.G.; Hack, A.; Minekus, M.; Zeijdner, E.; Cornelis, C.; Schoeters, G.; Verstraete, W.; Van de Wiele, T.; Wragg, J.; Rompelberg, C.J.M.; et al. Comparison of five in vitro digestion models to study the bioaccessibility of soil contaminants. *Environ. Sci. Technol.* **2002**, *36*, 3326–3334. [CrossRef] [PubMed]
6. ISO 17924:2018. In *Soil Quality—Assessment of Human Exposure from Ingestion of Soil and Soil Material—Procedure for the Estimation of the Human Bioaccessibility/Bioavailability of Metals in Soil*; ISO: Genève, Switzerland, 2018.
7. E1739-95 ASTM. In *Standard Guide for Risk Based Corrective Action Applied at Petroleum Release Sites*; ASTM International: West Conshohocken, PA, USA, 2015.
8. E2081-00 ASTM. In *Standard Guide for Risk Based Corrective Action*; ASTM International: West Conshohocken, PA, USA, 2015.
9. Ruby, M.V.; Fehling, K.A.; Paustenbach, D.J.; Landenberger, B.D.; Holsapple, M.P. Oral Bioaccessibility of Dioxins/Furans at Low Concentrations (50–350 ppt Toxicity Equivalent) in Soil. *Environ. Sci. Technol.* **2002**, *36*, 4905–4911. [CrossRef]
10. Ruby, M.V.; Davis, A.; Link, T.E.; Schoof, R.; Chaney, R.L.; Freeman, G.B.; Bergstrom, P. Development of an in vitro screening test to evaluate the in vivo bioaccessibility of ingested mine-waste lead. *Environ. Sci. Technol.* **1993**, *27*, 2870–2877. [CrossRef]
11. Ruby, M.V.; Schoof, R.; Brattin, W.; Goldade, M.; Post, G.; Harnois, M.; Mosby, D.E.; Casteel, S.W.; Berti, W.; Carpenter, M.; et al. Advances in evaluating the oral bioavailability of inorganics in soil for use in human health risk assessment. *Environ. Sci. Technol.* **1999**, *33*, 3697–3705. [CrossRef]
12. Davis, A.; Ruby, M.V.; Goad, P.; Eberle, S.; Chryssoulis, S. Mass balance on surface-bound mineralogic, and total lead concentrations as related to industrial aggregate bioaccessibility. *Environ. Sci. Technol.* **1997**, *31*, 37–44. [CrossRef]
13. Casteel, S.W.; Cowart, R.P.; Weis, C.P.; Henningsen, G.M.; Hoffman, E.; Brattin, W.J.; Guzman, R.E.; Starost, M.F.; Payne, J.T.; Stockham, S.L.; et al. Bioavailability of lead to juvenile swine dosed with soil from the Smuggler Mountain NPL Site of Aspen, Colorado. *Toxicol. Sci.* **1997**, *36*, 177–187. [CrossRef]

14. Freeman, G.B.; Johnson, J.D.; Killinger, J.M.; Liao, S.C.; Feder, P.I.; Davis, A.O.; Ruby, M.V.; Chaney, R.L.; Lovre, S.C.; Bergstrom, P.D. Relative bioavailability of lead from mining waste soil in rats. *Fundam. Appl. Toxicol.* **1992**, *19*, 388–398. [CrossRef]
15. Wragg, J.; Cave, M.R. *In-Vitro Methods for the Measurement of the Oral Bioaccessibility of Selected Metals and Metalloids in Soils: A Critical Review*; Environment Agency: Bristol, UK, 2003.
16. Medlin, E.A. An in vitro method for estimating the relative bioavailability of lead in humans. Master's Thesis, Department of Geological Sciences, University of Colorado at Boulder, Boulder, CO, USA, 1997.
17. U.S. EPA. *Standard Operating Procedure for an In Vitro Bioaccessibility Assay for Lead and Arsenic in Soil*; OLEM 9200.2-164; U.S. EPA: Washington, DC, USA, 2017.
18. Rodriguez, R.R.; Basta, N.T. An in vitro gastrointestinal method to estimate bioavailable arsenic in contaminated soils and solid media. *Environ. Sci. Technol.* **1999**, *33*, 642–649. [CrossRef]
19. DIN E 19738. In *Deutsches Institut fur Normung e.V. Soil Quality—Absorption Availability of Organic and Inorganic Pollutants from Contaminated Soil Material*. 2000. Available online: https://www.techstreet.com/standards/din-19738?product_id=1982137#document (accessed on 22 March 2021).
20. Oomen, A.G.; Rompelberg, C.J.M.; Bruil, M.A.; Dobbe, C.J.G.; Pereboom, D.P.K.H.; Sips, A.J.A.M. Development of an In Vitro Digestion Model for Estimating the Bioaccessibility of Soil Contaminants. *Arch. Environ. Contam. Toxicol.* **2003**, *44*, 281–287. [CrossRef]
21. Juhasz, A.L.; Weber, J.; Smith, E.; Naidu, R.; Rees, M.; Rofe, A.; Kuchel, T.; Sansom, L. Assessment of four commonly employed in vitro arsenic bioaccessibility assays for predicting in vivo relative arsenic bioavailability in contaminated soils. *Environ. Sci. Technol.* **2009**, *43*, 9487–9494. [CrossRef]
22. Denys, S.; Caboche, J.; Tack, K.; Rychen, G.; Wragg, J.; Cave, M.; Jondreville, C.; Feidt, C. In vivo validation of the unified BARGE method to assess the bioaccessibility of arsenic, antimony, cadmium, and lead in soils. *Environ. Sci. Technol.* **2012**, *46*, 6252–6260. [CrossRef]
23. Wragg, J.; Cave, M.; Basta, N.; Brandon, E.; Casteel, S.; Denys, S.; Gron, C.; Oomen, A.; Reimer, K.; Tack, K.; et al. An inter-laboratory trial of the unified BARGE bioaccessibility method for arsenic, cadmium and lead in soil. *Sci. Total Environ.* **2011**, *409*, 4016–4030. [CrossRef]
24. Drexler, J.W.; Brattin, W.J. An in vitro procedure for estimation of lead relative bioavailability: With validation. *Hum. Ecol. Risk Assess.* **2007**, *13*, 383–401. [CrossRef]
25. USEPA. *Validation Assessment of In Vitro Lead Bioaccessibility Assay for Predicting Relative Bioavailability of Lead in Soils and Soil-Like Materials at Superfund Sites*; OSWER 9200; U.S. Environmental Protection Agency: Washington, DC, USA, 2009.
26. Izquierdo, M.; De Miguel, E.; Ortega, M.F.; Mingot, J. Bioaccessibility of metals and human health risk assessment in community urban gardens. *Chemosphere* **2015**, *135*, 312–318. [CrossRef]
27. Mendoza, C.J.; Garrido, R.T.; Quilodrán, R.C.; Segovia, C.M.; Parada, A.J. Evaluation of the bioaccessible gastric and intestinal fractions of heavy metals in contaminated soils by means of a simple bioaccessibility extraction test. *Chemosphere* **2017**, *176*, 81–88. [CrossRef] [PubMed]
28. Guney, M.; Zagury, G.J.; Dogan, N.; Onay, T.T. Exposure assessment and risk characterization from trace elements following soil ingestion by children exposed to playgrounds, parks and picnic areas. *J. Hazard. Mater.* **2010**, *182*, 656–664. [CrossRef] [PubMed]
29. Mehta, N.; Cipullo, S.; Cocerva, T.; Coulon, F.; Dino, G.A.; Ajmone-Marsan, F.; Padoan, E.; Cox, S.F.; Cave, M.R.; De Luca, D.A. Incorporating oral bioaccessibility into human health risk assessment due to potentially toxic elements in extractive waste and contaminated soils from an abandoned mine site. *Chemosphere* **2020**, *255*, 126927. [CrossRef] [PubMed]
30. Zhang, R.; Han, D.; Jiang, L.; Zhong, M.; Liang, J.; Xia, T.X.; Zhao, Y. Derivation of site-specific remediation goals by incorporating the bioaccessibility of polycyclic aromatic hydrocarbons with the probabilistic analysis method. *J. Hazard. Mater.* **2020**, *384*, 121239. [CrossRef]
31. Ma, J.; Li, Y.; Liu, Y.; Lin, C.; Cheng, H. Effects of soil particle size on metal bioaccessibility and health risk assessment. *Ecotoxicol. Environ. Saf.* **2019**, *186*, 109748. [CrossRef]
32. Li, Y.; Padoan, E.; Ajmone-Marsan, F. Soil particle size fraction and potentially toxic elements bioaccessibility: A review. *Ecotoxicol. Environ. Saf.* **2021**, *209*, 111806. [CrossRef]
33. Urrutia-Goyes, R.; Argyraki, A.; Ornelas-Soto, N. Assessing Lead, Nickel, and Zinc Pollution in Topsoil from a Historic Shooting Range Rehabilitated into a Public Urban Park. *Int. J. Environ. Res. Public Health* **2017**, *14*, 698. [CrossRef]
34. Fayiga, A.O.; Saha, U.K. Soil pollution at outdoor shooting ranges: Health effects, bioavailability and best management practices. *Environ. Pollut.* **2016**, *216*, 135–145. [CrossRef]
35. Walraven, N.; Bakker, M.; van Os, B.J.H.; Klaver, G.T.; Middelburg, J.J.; Davies, G.R. Factors controlling the oral bioaccessibility of anthropogenic Pb in polluted soils. *Sci. Total Environ.* **2015**, *506–507*, 149–163. [CrossRef] [PubMed]
36. Sanderson, P.; Naidu, R.; Bolan, N.; Bowman, N.; Mclure, S. Effect of soil type on distribution and bioaccessibility of metal contaminants in shooting range soils. *Sci. Total Environ.* **2012**, *438*, 452–462. [CrossRef] [PubMed]
37. Smith, E.; Weber, J.; Naidu, R.; McLaren, R.G.; Juhasz, A.L. Assessment of lead bioaccessibility in peri-urban contaminated soils. *J. Hazard. Mater.* **2011**, *186*, 300–305. [CrossRef] [PubMed]
38. Moseley, R.A.; Barnett, M.O.; Stewart, M.A.; Mehlhorn, T.L.; Jardine, P.M.; Ginder-Vogel, M.; Fendorf, S. Decreasing lead bioaccessibility in industrial and firing range soils with phosphate-based amendments. *J. Environ. Qual.* **2008**, *37*, 2116–2124. [CrossRef]

39. Cao, X.; Ma, L.Q.; Chen, M.; Hardison, D.W.; Harris, W.G. Weathering of lead bullets and their environmental effects at outdoor shooting ranges. *J. Environ. Qual.* **2003**, *307*, 526–534. [CrossRef]
40. Sanderson, P.; Qi, F.; Seshadri, B.; Wijayawardena, A.; Naidu, R. Contamination, fate and management of metals in shooting range soils—a review. *Curr. Pollut. Rep.* **2018**, *4*, 175–187. [CrossRef]
41. U.S. EPA. *Method 3051A: Microwave Assisted acid Digestiom of Sediments, Sludges, Soils and Oils*; U.S. EPA: Washington, DC, USA, 2007.
42. U.S. EPA. *Recommendations for Sieving Soil and Dust Samples at Lead Sites for Assessment of Incidental Ingestion*; OLEM 9200.1-129; U.S. EPA: Washington, DC, USA, 2016.
43. Ruby, M.V.; Lowney, Y.W. Selective soil particle adherence to hands: Implications for understanding oral exposure to soil contaminants. *Environ. Sci. Technol.* **2012**, *46*, 12759–12771. [CrossRef] [PubMed]
44. APAT-ISPRA. Criteri Metodologici per l'Applicazione dell'Analisi Assoluta di Rischio ai Siti contaminati. 2008. Available online: https://www.isprambiente.gov.it/it/attivita/suolo-e-territorio/siti-contaminati/analisi-di-rischio (accessed on 24 January 2021).
45. Legislative Decree No. 152/06. In *Environmental Framework Regulation*; Italian Official Bulletin No. 88; Italian Official Bulletin: Roma, Italy, 2006.
46. Juhasz, A.L.; Smith, E.; Weber, J.; Rees, M.; Kuchel, T.; Rofe, A.; Sansom, L.; Naidu, R. Predicting lead relative bioavailability in peri-urban contaminated soils using in vitro bioaccessibility assays. *J. Environ. Sci. Health Part A* **2013**, *48*, 604–611. [CrossRef] [PubMed]

Article

Enhanced Electrokinetic Remediation for the Removal of Heavy Metals from Contaminated Soils

Claudio Cameselle *, Susana Gouveia and Adrian Cabo

Department of Chemical Engineering, BiotecnIA, University of Vigo, 36310 Vigo, Spain; gouveia@uvigo.es (S.G.); acabo@uvigo.es (A.C.)
* Correspondence: claudio@uvigo.es; Tel.: +34-986-812318

Abstract: The electrokinetic remediation of an agricultural soil contaminated with heavy metals was studied using organic acids as facilitating agents. The unenhanced electrokinetic treatment using deionized water as processing fluid did not show any significant mobilization and removal of heavy metals due to the low solubilization of metals and precipitation at high pH conditions close to the cathode. EDTA and citric acid 0.1 M were used as facilitating agents to favor the dissolution and transportation of metals. The organic acids were added to the catholyte and penetrated into the soil specimen by electromigration. EDTA formed negatively charged complexes. Citric acid formed neutral metal complexes in the soil pH conditions (pH = 2–4). Citric acid was much more effective in the dissolution and transportation out of the soil specimen of complexed metals. In order to enhance the removal of metals, the concentration of citric acid was increased up to 0.5 M, resulting in the removal of 78.7% of Cd, 78.6% of Co, 72.5% of Cu, 73.3% of Zn, 11.8% of Cr and 9.8% of Pb.

Keywords: contaminated soil; metal; citric acid; EDTA; electrokinetic remediation

Citation: Cameselle, C.; Gouveia, S.; Cabo, A. Enhanced Electrokinetic Remediation for the Removal of Heavy Metals from Contaminated Soils. *Appl. Sci.* **2021**, *11*, 1799. https://doi.org/10.3390/app11041799

Academic Editor: Fulvia Chiampo
Received: 3 February 2021
Accepted: 16 February 2021
Published: 18 February 2021

Publisher's Note: MDPI stays neutral with regard to jurisdictional claims in published maps and institutional affiliations.

Copyright: © 2021 by the authors. Licensee MDPI, Basel, Switzerland. This article is an open access article distributed under the terms and conditions of the Creative Commons Attribution (CC BY) license (https://creativecommons.org/licenses/by/4.0/).

1. Introduction

Soil is affected by various anthropogenic activities, including mining, waste dump, deforestation, urbanization, intensive agriculture practices and change in land use [1]. As a result, soils have been contaminated with a variety of organic and inorganic contaminants [2]. The metal-contamination in the soil is of special concern because metals show special toxicity for living organisms affecting the development of microorganisms and plants, enter the food change and extend the toxicity to animals and humans [3]. Contaminating metals cannot be degraded, and their removal from the soil is usually difficult, time-consuming and costly [4]. Many studies published in literature dealt with the removal of heavy metals from soil. However, the efforts in developing and testing various innovative technologies did not still result in a reliable and feasible technology to be generally applied in the remediation of metal-contaminated soils [5,6]. Recently, research has focused on the development of high-sensitivity sensors, usually based on electrochemical technology [7,8]. These sensors can be used for the detection of heavy metal ions in various media, allowing a new way to monitoring metal-contamination in hard-to-reach locations and remote environments.

Electrokinetic remediation has been proposed as an effective technology for the remediation of metal-contaminated soils [9]. Electrokinetic technology uses a low-intensity DC electric field to mobilize and transport the metals out of the soil. The benefits of electrokinetics rely on its capacity to be applied in situ with minimum disturbing of the surface, the possibility of treating low permeability soils and the low-cost in energy. However, electrokinetics is affected by the geochemical soil conditions that may limit the solubilization of metals. In order to increase the solubility of the metal in the soil interstitial fluid, various methods have been proposed [10]. The acidification of soil with the acid front electrogenerated in the anode is the most common option [11]. This option is usually combined with the depolarization of the electro-reduction of water on the cathode to suppress the alkaline front [12]. The controlled addition of mineral acid in the cathode allows for the neutralization of the basic front and favors the acidification of the soil with the subsequent solubilization of metals (for example, Co^{2+}, Ni^{2+}, Cd^{2+}, ...). Alternatively, alkaline metals (for example, chromate) can be extracted from the soil matrix favoring the development of the basic front on the cathode and suppressing the acid front on the anode by the addition of alkali (e.g., NaOH). The modification of the soil pH with the acid or alkali front from cathode or anode is an effective method for the solubilization of metals in many cases, but it is important to consider the geochemistry of soils and contaminating metals. As an example, Ottosen et al. [13] used the controlled addition of ammonia to solubilize copper and arsenic. Ammonia forms a stable complex with copper $[Cu(NH_3)_4]^{2-}$ in alkaline conditions that can be transported towards the cathode by electromigration. Conversely, various authors [12,14] used sulfuric acid to neutralize the basic environment of the cathode to mobilize the metals due to the acidification of the acid front from the anode. This method was very effective for metals, such as Cd^{2+}, Co^{2+} and Zn^{2+}, but Pb was not removed because it forms an insoluble salt with sulfate [14]. These results proved that the chemistry of facilitating agents with the metal–soil system is of crucial importance for a successful remediation. Furthermore, the acidification or alkalinization of soil may induce unacceptable changes in soil properties. The soil will require additional treatment to restore the natural properties of soil at the end of the remediation process [9].

The use of organic acids as facilitating agents in electrokinetics shows various benefits for the remediation of metal-contaminated soils. Organic acids are recognized as effective complexing and chelating agents that can keep metals in solution in a wide range of pH, even in an alkaline environment [15]. Moreover, the mild conditions of organic acids do not induce dramatic changes in the soil properties. Organic acids are usually biodegradable and can be removed by natural biodegradation at the end of the remediation process. Furthermore, the residual organic acids in the soil may help to enhance the activity of the natural soil microflora.

This paper studies the remediation of six metals (Cd, Cr, Co, Cu, Pb, and Zn) from agricultural soil with aged contamination. The different chemical nature of the metal species and the presence of these metals in high concentrations make the remediation of soil more difficult. The removal of metals was studied using organic acids (EDTA and citric acid) as facilitating agents with the objective to develop a feasible technology to be applied at a large scale. The aim of the study is to identify the facilitating agent and the experimental conditions to achieve the mobilization and removal of the six metallic species from the soil.

2. Materials and Methods

2.1. Soil Characteristics and Contamination Procedure

The characteristics of soil used in this study are listed in Table 1. The soil was sampled on a farm close to the campus of the University of Vigo in the NW of Spain. The soil was sampled from the upper layer of soil (0–20 cm depth) after removing the vegetative cover. The soil was extended in a thin layer (2–5 cm) in the lab to let it dry for 72 h to a moisture content of 5% (wet basis). Then, it was sieved through a 2 cm mesh to remove small stones, sticks, roots, and other non-soil components. The soil specimen was thoroughly mixed to obtain a uniform sample that was contaminated with 6 heavy metals: Cd, Cr, Co, Cu, Pb and Zn, using the necessary amounts of the corresponding salts to achieve the spiking concentration in Table 2. After the spiking process, the soil was stored in a plastic box in the darkness for 2 years to ensure the aging of the contaminants. The objective of this process was to ensure the behavior of the contaminated soil similar to the behavior of real contaminated soil.

Table 1. Characteristics of soil used in the experiments.

Parameter	Value	Testing Method
pH	4.4	ASTM D1293 [16]
Electrical conductivity (mS/cm)	3.15	ASTM D1293 [16]
Specific gravity (dry basis, g/cm^3)	2.7	ASTM D 854 [17]
Water holding capacity	48.5%	ASTM D 2980 [18]
Moisture content (wet basis)	22.4%	ASTM D2216 [19]
Soil organic content (dry basis)	4.5%	ASTM D 2974 [20]
Particle size analysis		ASTM D 422 [21]
Clay (<0.002 mm)	44.3%	
Silt (0.002–0.05 mm)	43.2%	
Sand (0.05–2 mm)	12.5%	

Table 2. Metal concentration in the contaminated soil.

Metal Species	Metal Salt	Spiking Concentration (mg/kg)
Cd^{2+}	$Cd(NO_3)_2\ 4H_2O$	141
Cr(VI)	$K_2Cr_2O_7$	1000
Co^{2+}	$CoCl_2\ 6H_2O$	185
Cu^{2+}	$CuSO_4\ 5H_2O$	1023
Pb^{2+}	$PbCl_2$	1000
Zn^{2+}	$ZnSO_4\ 7H_2O$	1001

2.2. Extraction Tests

The capacity of 9 extracting agents and water to dissolve the metals from soil was tested. The extraction tests used 2 g of soil (dry) and 100 mL of the extracting solution. Four mineral acids (hydrochloric, nitric, phosphoric and sulfuric acid), five organic acids (acetic, citric, EDTA, oxalic and tartaric acid) and deionized water were used as extracting solutions. The concentration of all the acid solutions was 0.1 M. The mixtures were shaken for 24 h at 180 rpm and 20 °C (±0.5 °C) in an orbital shaker. The metals were determined by ICP-OES (Inductively coupled plasma–optical emission spectrometry) in the supernatant solution.

2.3. Electrokinetic Setup and Testing

Figure 1 shows the electrokinetic cell used in this study. The cell was composed of a central tube (20 cm long and 4 cm diameter) that holds the soil specimen. 320 g of contaminated soil was used in each test. The soil specimen was introduced manually in the central tube, and it was compacted with a plastic piston. The central tube was connected to two electrode chambers filled with the processing fluid selected in each test (deionized water, citric acid, EDTA or NaOH), as shown in Table 3. The volume of the electrode chambers was 300 mL, and the total volume of the processing fluid, including the electrode chamber and the expansion vessel, was 500 mL. The soil specimen was separated into the processing fluid by a porous stone and filter paper. The electrodes were installed in the electrode chambers immersed in the processing fluid and connected to a DC power supply. All the tests were conducted at constant DC electric potential (20 V) for 30 days, except test 4 with citric acid 0.5 M in anolyte and catholyte that was run for 65 days. The anode was made of titanium, and the cathode was made of stainless steel.

At the end of the tests, the electrode chambers were emptied, and the processing fluid was collected in plastic bottles and stored in the fridge at 4 °C until analysis. The soil was extracted from the central tube and divided into 6 equal sections. Solid and liquid samples were digested with nitric acid and hydrochloric acid as per the USEPA method 3010A (acid digestion of aqueous samples and extracts for total metals for analysis by FLAA or ICP spectroscopy) [22] or method 3050B (acid digestion of sediments, sludges, and soils) [23]. Metal concentrations were determined by ICP-OES in the supernatant solution of the acid digestions. Soil pH was determined by the ASTM D1293 [16] method using 1 g of soil and

2.5 mL of 1 M KCl solution. The mixture was shaken for 1 h, and the pH was measured in the supernatant fluid.

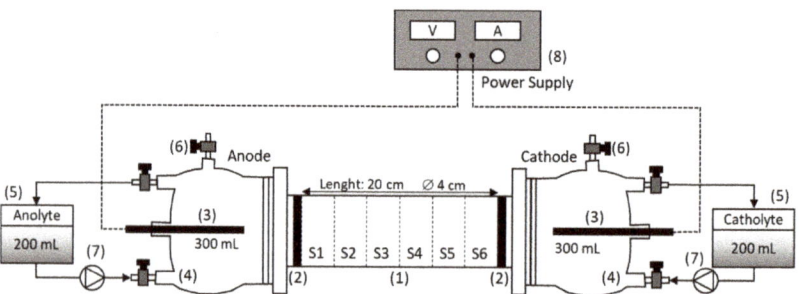

Figure 1. Experimental setup. (1) Soil specimen, soil sections are represented as S1–S6, (2) filter paper and porous stone, (3) electrodes (anode and cathode), (4) electrode chambers, (5) processing fluid vessel (anolyte and catholyte), (6) gas release valves, (7) recirculating pumps, (8) DC power supply.

Table 3. Experimental conditions of the electrokinetic tests.

Test	Anolyte	Soil	Catholyte	Potential Drop	Time
Test 1	Deionized water	320 g	DI water	20 V	30 days
Test 2	Deionized water	320 g	0.1 M citric acid	20 V	30 days
Test 3	0.1 M NaOH	320 g	0.1 M EDTA	20 V	30 days
Test 4	0.5 M Citric acid	320 g	0.5 M citric acid	20 V	65 days

3. Results

3.1. Extraction of Metals from the Soil

Figure 2 shows the metal extraction fraction when 2 g of soil was in contact for 24 h with 100 mL of 0.1 M solution of the corresponding extracting agent. In general, Cd, Co, Cu and Zn were easier dissolved than Pb and Cr, with both mineral acid and organic acids. The mineral acid solutions tended to show slightly better extraction ratios than organic acids. This is due to the lower pH solution of the mineral acids that extracted the metals from the surface of the soil particles by the direct attack of the H^+ ions, stabilizing the metal in solution at pH below 1. Conversely, the organic acids showed a solubilization ratio based on their capacity to form soluble complexes and chelates in solution [24]. A closer analysis of the extraction results in Figure 2 revealed that the metal solubilization ratio depended on the chemistry of each pair of metal–acid. Hydrochloric acid and nitric acid showed 70–80% extraction for all the metals, but Cr, because this metal was present in the soil as Cr(VI) and the acidification was not able to desorb the anionic Cr species. Phosphoric and sulfuric acid also showed a high extraction ratio for Cd, Co, Cu and Zn, but very low extraction for lead because of the formation of insoluble salts between Pb and sulfate ($PbSO_4$, Ksp = 1.6×10^{-8}) or phosphate ($Pb_3(PO_4)_2$, Ksp = 9.9×10^{-55}). The organic acids showed a 60–70% extraction ratio for Cd, Co, Cu and Zn. The lead was extracted with EDTA, oxalic acid and tartaric acid, whereas acetic and citric acids showed negligible removal due to the tendency of lead to adsorb to the soil particles and organic matter [25]. Cr was extracted with oxalic acid due to the low acidity of the solution (pH = 1.40) and the reducing capacity of oxalic that transformed Cr(VI) into Cr(III) and formed stable complexes in solution [26].

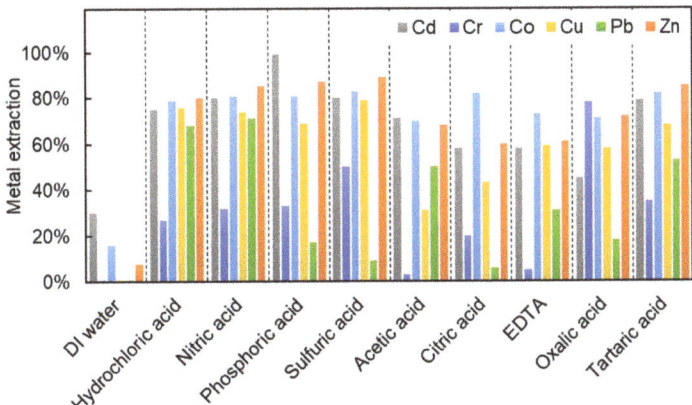

Figure 2. Fraction of extracted metals from the soil with mineral acids, organic acids and water (All the compounds were tested at 0.1 M).

The information in Figure 2 can be used for the selection of facilitating agents for the electrokinetic remediation of the contaminated soil. Mineral acids were very effective in the mobilization of the metals, but their activity was always associated with a very acidic pH, below 1. In the remediation of soils, such low pH will affect the physicochemical and biological properties of soil to an unacceptable level, and further treatment would be necessary to restore the natural properties of soil. The organic acids seem to be more appropriate for soil remediation projects because they were able to extract the contaminated metals in less harsh conditions than mineral acids. Moreover, the organic acids tested in Figure 2 are biodegradable and can be removed from the soil for natural biotic degradation while enhancing the microbial activity of the soil [27]. Considering the results in Figure 2 and the available information in the literature, EDTA and citric acid were selected as appropriate facilitating agents for the electrokinetic remediation of soil [14,28]. EDTA is a complexing agent for all the contaminating metals in the soil specimen [24]. EDTA can be degraded by the soil microflora at the end of the electrokinetic treatment [29]. Citric acid is recognized as a non-toxic compound since it is present in all the eukaryotic cells in the Krebs cycle (the cycle of the tricarboxylic acids). Moreover, citric acid showed a special activity in the electrokinetic remediation of soils contaminated with metals [14].

3.2. Electrokinetic Treatment

The removal of heavy metals from the soil specimen was analyzed in a set of electrokinetic tests using EDTA and citric acid as facilitating agents to enhance the solubilization of the metals and their removal in the processing fluid in the electrode chambers. The results of the tests with EDTA and citric acid were compared with an unenhanced electrokinetic test using deionized water as processing fluid in anode and cathode (test 1). Test 2 used 0.1 M citric acid as catholyte and deionized water as anolyte. Test 3 used 0.1 M EDTA as catholyte and 0.1 M NaOH as anolyte. Finally, an additional electrokinetic test (test 4) was run with 0.5 M citric acid as processing fluid in anode and cathode with the objective to enhance the solubilization and removal of metals from soil (Table 3).

Figure 3 shows the pH in the electrode chambers and the pH profile in the soil for tests 1, 2, 3 and 4. The unenhanced electrokinetic test with deionized water as processing fluid (test 1) showed the typical pH profile induced by the electrolysis of water upon the electrodes [9]. The pH decreases in the anode to pH 2 and increases in the cathode to pH higher than 12. The pH of the soil in test 1 (the blue line in Figure 3) decreases in the fraction of soil close to the anode and increases in the fraction of soil close to the cathode. The profile of soil pH has an enormous influence on the speciation and mobilization of the contaminating metals in soil (Figure 4). Cd, Cu and Zn were mobilized in sections 1–3, where the soil pH was lower than pH = 4 (the initial pH of soil). These metals were accumulated in sections 4–6, where the pH conditions induced the precipitation of metals in the soil forming the corresponding metallic hydroxide, especially in sections 5 and 6. Cobalt was mobilized in section 1–4 and mainly accumulated in section 6 because the immobilization of cobalt as $Co(OH)_2$ takes place at pH > 7. Chromium and lead were not mobilized or removed in the unenhanced electrokinetic test. Chromium is an anionic metal, and it was not expected to be mobilized in very acid conditions. In fact, a slight mobilization of Cr was observed in section 6, where the pH was alkaline. Lead is a metal that can be retained in the soil attached to the organic matter, and it is difficult to mobilize [25]. The alkaline environment in section 6 may have favored the dissolution of organic matter and a slight mobilization of Pb (Figure 4). Overall, the unenhanced electrokinetic test was not effective in the removal of metals from soil. Four metal species were redistributed in the soil (Cd, Co, Cu and Zn), and two species (Cr and Pb) showed no significant mobilization. No significant metal removal was observed, as presented in Table 4. Test 2 and 3 were designed to evaluate the efficiency of citric acid and EDTA to mobilize and remove the metals from soil.

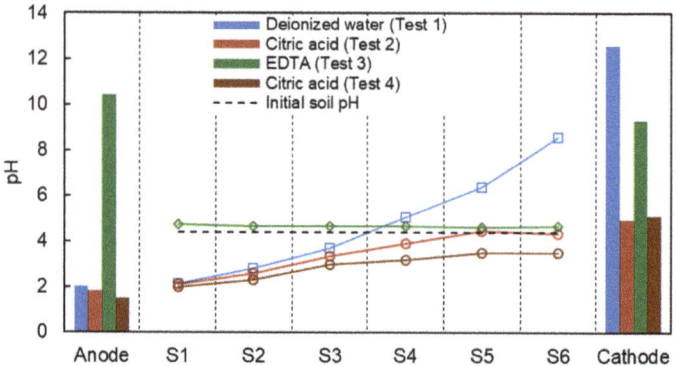

Figure 3. pH in soil and processing fluid in the electrode chambers for the electrokinetic tests with deionized water (test 1), 0.1 M citric acid (test 2), 0.1 M EDTA (test 3) and 0.5 M citric acid (test 4).

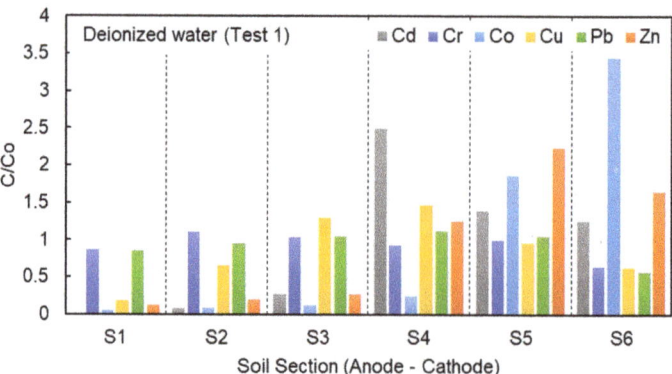

Figure 4. Unenhanced electrokinetic test of soil contaminated with metals using deionized water as processing fluid in anode and cathode chambers. C is the metal concentration in soil at the end of the test, and Co is the initial metal concentration in soil.

Table 4. Mass balance of the metals in the soil at the end of the electrokinetic tests.

Test—Catholyte	Fractions	Cd	Cr	Co	Cu	Pb	Zn
Test 1—Deionized water	Soil (%)	98.1%	97.4%	99.9%	97.4%	96.8%	98.5%
	Anode (%)	0.0%	0.5%	0.1%	0.0%	0.0%	0.0%
	Cathode (%)	0.0%	0.0%	0.0%	0.0%	0.0%	0.0%
	Error (%)	1.9%	2.1%	0.1%	2.6%	3.2%	1.5%
Test 2—0.1 M citric acid	Soil (%)	96.1%	95.4%	92.7%	93.5%	97.8%	86.7%
	Anode (%)	0.1%	0.1%	0.1%	0.0%	0.0%	0.1%
	Cathode (%)	0.2%	0.0%	4.5%	2.1%	0.0%	9.3%
	Error (%)	3.7%	4.5%	2.7%	4.3%	2.1%	3.9%
Test 3—0.1 M EDTA	Soil (%)	99.7%	99.8%	98.3%	94.1%	99.4%	99.2%
	Anode (%)	0.0%	0.0%	0.1%	0.0%	0.0%	0.1%
	Cathode (%)	0.6%	0.1%	0.9%	0.2%	0.3%	0.9%
	Error (%)	−0.3%	0.1%	0.7%	5.6%	0.3%	−0.1%
Test 4—0.5 M citric acid	Soil (%)	21.3%	88.2%	21.4%	27.5%	90.2%	26.7%
	Anode (%)	1.2%	2.3%	0.3%	0.0%	0.0%	0.1%
	Cathode (%)	78.3%	6.3%	76.9%	71.4%	8.2%	73.9%
	Error (%)	−0.8%	3.2%	1.4%	1.1%	1.6%	−0.7%

Test 2 used citric acid as a facilitating agent in the cathode. The citrate ion penetrated the soil specimen by electromigration, mobilizing the metals that could be stabilized in solution forming complexes or chelates with citrate. The presence of citric acid in the cathode avoided the formation of an alkaline environment. The pH in the cathode at the end of the test was pH = 5 that was the result of the partial neutralization of the 0.1 M citric acid solution (the natural pH of the citric acid solution was 2) with the electrochemical consumption of H^+ on the cathode. The anolyte was deionized water. The electrolysis of water in the anode produced H^+ ions that formed an acid front that acidified the soil specimen. As a result, the soil pH decreased in sections 1–4 and remained at the initial soil pH in sections 5 and 6 (Figure 3). The pH profile in the soil with the combination of the citrate as a complexing agent in the interstitial fluid showed a decisive influence in the speciation of metals and their transport along the soil specimen. As presented in Figure 5, Zn was mobilized in sections 1–3 and accumulated in sections 4 and 5. The concentration of Zn in the last section 6 remained unchanged. This behavior can be explained with a closer analysis of the pH in the soil and the chemical speciation of the system Zn^{2+}-citrate (Figure 6). Zn^{2+} is the predominant species in the solution at a very acid pH < 1.5. In the presence of citrate, Zn^{2+} forms six different complexes (negative, neutral and positive complexes). The relative predominance of the complexes depends on the medium pH. At pH = 2–3 the dominant species is the positive complex $[ZnH_2L]^+$. L represents the ligand citrate. At pH = 3–4.5 the dominant species is the neutral complex [ZnHL]. At pH > 5, the dominant species is the negative complex $[ZnL]^-$. The pH in the soil sections 1–4 was below pH = 4, so Zn was mobilized by the combined effect of the H^+ attack and the stabilization in the solution of Zn^{2+} with citrate forming positive or neutral complexes. These complexes were transported towards the cathode by electromigration and/or electroosmosis. In soil sections 5 and 6, the pH is about 5, and the predominant species was the negative complex $[ZnL]^-$. The negative complex will be transported towards the anode by electromigration. This explains why Zn is accumulated in section 5 of the soil specimen. Similar behavior was found for Cd, Cu, Co. These three metals form various neutral, positive and negative complexes with citrate [24], similar to the case explained for Zn. As a result, Cd was mobilized in sections 1–3 and 6 and accumulated in sections 4 and 5. Cu was removed from both ends of the soil specimen due to the formation of positive complexes on the anode side and negative complexes on the cathode side. Cu was accumulated in the center of the soil specimen (sections 3 and 4). In a similar way, cobalt was accumulated in sections 5 and 6. It is noteworthy the small but significant mobilization of Pb in the anode and cathode side and a slight accumulation in the center of the soil specimen. These results confirm that Pb can be dissolved with citric acid, whereas it was not mobilized in test 1 with deionized water. Finally, chromium was mobilized on the cathode side by the combined effect of the pH and the presence of citrate. Cr was accumulated in soil sections 2–4. As a conclusion, citrate appeared to be an active compound for the mobilization of all the metals, including Cr and Pb, that showed major resistance to the mobilization in the unenhanced electrokinetic test 1. However, there was minor net removal of metals from soil in the electrode solutions. Only 9% of Zn, 4.5% of Co and 2% of Cu were found in the cathode chamber (Table 4). These results suggested that the treatment time and test conditions (pH, concentration of citric acid, etc.) were not enough for the effective removal of metals from contaminated soil.

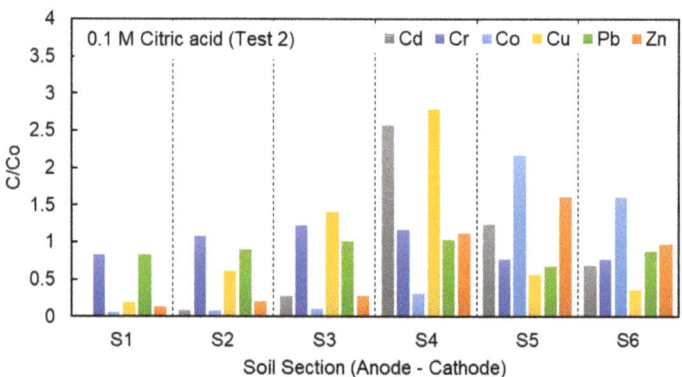

Figure 5. Electrokinetic test of soil contaminated with metals using 0.1 M citric acid as catholyte and deionized water as anolyte. C is the metal concentration in soil at the end of the test, and Co is the initial metal concentration in soil.

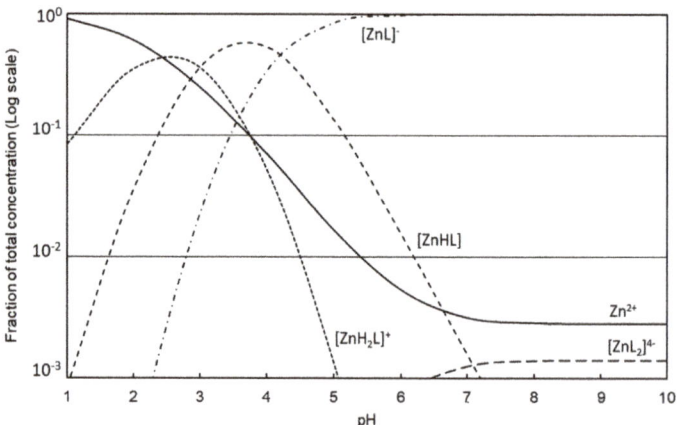

Figure 6. Speciation of zinc and zinc complexes formation in an aqueous system with 0.1 M citrate and 0.1 M Zn^{2+}. (L represents the ligand citrate).

Test 3 was run with EDTA 0.1 M as catholyte. The electrokinetic phenomena would transport EDTA into the soil specimen by electromigration. EDTA forms complexes and chelates with all the metals [24] in the soil specimen. The analysis of the Zn^{2+} speciation with EDTA in an aqueous system is shown in Figure 7. EDTA formed three different complexes with EDTA formed three complexes with EDTA: $[ZnL]^{2-}$, $[ZnHL]^{-}$, $[ZnOHL]^{3-}$. The relative abundance of each complex depends on the pH. EDTA is very effective in complexing Zn in the studied pH range. The relative abundance of the cationic species

Zn^{2+} was always below 0.1% (<10^{-3} in Figure 7) at any pH. That is the reason the cationic Zn^{2+} does not appear in Figure 7. The speciation analysis informs that EDTA formed a very stable negative complex $[ZnL]^{2-}$ with zinc in all the pH intervals (pH = 3.5–10). The study did not include pH below 3.5 because EDTA is not stable in solution in acidic pH. In conclusion, the use of EDTA requires neutral or slightly alkaline pH to favor the solubility of EDTA and the stability of the metal complexes. Hence, in test 3, the anolyte solution was 0.1 M NaOH to neutralize the acid environment generated by the electrolysis of water in the anode. Figure 3 shows the pH of test 3. The electrode solutions were alkaline as expected due to the electrolysis of water in the cathode and the use of NaOH in the anode. The penetration in the soil of the alkaline front from the cathode slightly increased the soil pH to 5. In these conditions, EDTA may form stable negative complexes with all the metals that would be transported towards the anode by electromigration. Figure 8 shows the residual metal in the soil at the end of test 3. EDTA was very effective in mobilizing Cd, Co, Cu, Pb and Zn, forming stable negative complexes that electromigrated towards the anode. However, in the time span of test 3, only the metal concentration decreased in section 6. The mobilized metals migrated towards the anode, confirming the formation of negative complexes, and accumulated in section 5. No significant changes in metal concentration were detected in sections 1–4. Cr did not show any mobilization because the anionic nature of this metal did not undergo any mobilization with EDTA at pH = 5. There was no net metal removal from the soil, as is shown in Table 4.

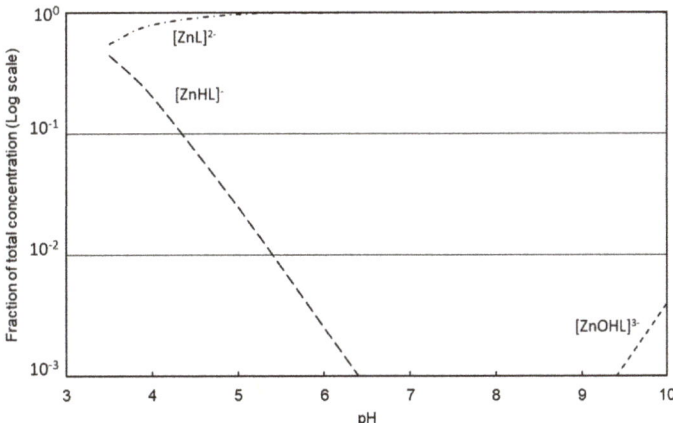

Figure 7. Speciation of zinc complexes in an aqueous system with 0.1 M EDTA and 0.1 M Zn^{2+}. (L represents the ligand EDTA).

Citric acid showed better mobilization of the metals than test 1 (with deionized water) and test 3 (with EDTA). Test 2 with citric acid showed a good mobilization of Cd, Co, Cu and Zn, but the metals were redistributed in the soil specimen with minor removal or accumulation in the electrode chambers (Table 4). It was hypothesized that the electrokinetic treatment with higher citric acid concentration and/or longer treatment time would increase the removal of metals from soil. Test 4 was designed to evaluate the capacity of citric acid to achieve an effective removal of the metals for the contaminated soil. Test 4 was operated with citric acid 0.5 M on both electrode chambers. Citric acid can be introduced into the soil specimen by electromigration from the cathode and by electroosmosis from the anode. It was expected that the most effective transportation of citric acid would correspond to the electromigration from the cathode, but the presence of citric acid on the anode would assure the presence of citrate in the fraction of soil close to the anode, favoring the solubilization of metals. This test was operated on for 65 days. Figure 9 shows the residual metal concentration in the soil after the electrokinetic test. The metal concentration profile for Cd, Co, Cu and Zn shows transportation towards the cathode.

The soil pH at the end of the test was below 3.5 (Figure 3). The pH conditions assure the mobilization of the metals by the combined effect of citrate and H⁺ ions, and the subsequent stabilization in solution was forming neutral or positive complexes (Figure 6). The metal complexes were transported towards the cathode by electromigration and electroosmosis. The global removal ratio at the end of the test was 78.7% of Cd, 78.6% of Co, 72.5% of Cu, and 73.3% of Zn. The lead was difficult to mobilize due to the strong adsorption of this metal to the organic matter and soil particles [25]. Only a small fraction of Pb (9.8%) was removed at the end of test 4. However, the profile of the residual Pb concentrations in soil suggests a mobilization of lead in sections 1–3 and its transportation towards the cathode. Chromium was removed by 11.8%. The profile of the residual concentrations in soil suggests that Cr was mobilized and transported towards the cathode. Considering that the contaminant Cr species was chromate, the transportation towards the cathode suggests a reduction of Chromate in the acid environment of the soil during the test by reaction with citrate or organic matter in the soil. Cr(III) was the resultant product of the chemical reduction of chromate. Cr(III) was complexed with citrate forming neutral or positive complexes that were transported towards the cathode [30]. The global mass balances for the six metals in this study (Table 4) confirmed the removal and transportation of the metals towards the cathode due to the complexing activity of citrate.

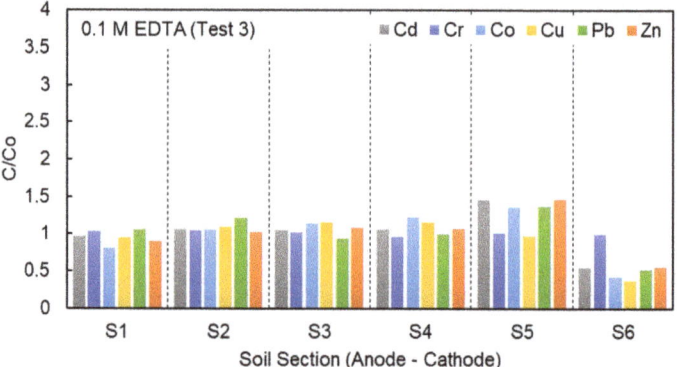

Figure 8. Electrokinetic test of soil contaminated with metals using 0.1 M EDTA as catholyte and 0.1 M NaOH as anolyte. C is the metal concentration in soil at the end of the test, and Co is the initial metal concentration in soil.

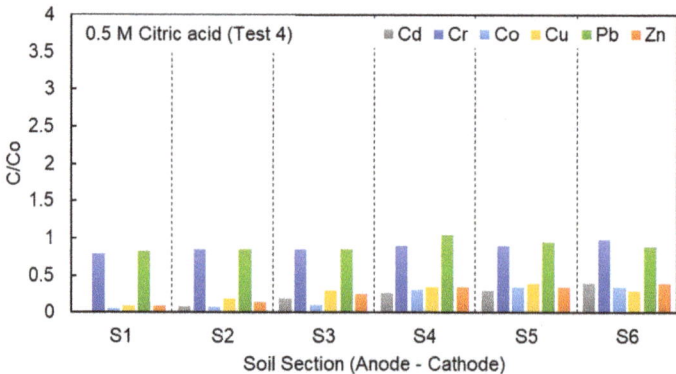

Figure 9. Electrokinetic test of soil contaminated with metals using 0.5 M citric acid as catholyte and anolyte. C is the metal concentration in soil at the end of the test, and Co is the initial metal concentration in soil.

4. Conclusions

Metal-contaminated soil is a serious problem, and electrokinetic remediation is a practical technology that can be used for their remediation. The success of electrokinetic remediation in removing metals depends on the experimental conditions and the selection of facilitating agents. This study proved that:

- Organic acids are a good alternative as facilitating agents because of the mild operating conditions and their capacity to dissolve and stabilize the metals in solution, forming stable complexes and chelates;
- EDTA 0.1 M was effective in the dissolution and complexing of aged contamination of Cd, Co, Cu, Pb and Zn. However, the removal rate was too slow to be considered for a large-scale application;
- Cr was not mobilized or removed in the electrokinetic treatment with EDTA;
- The use of citric acid 0.1 M as processing fluid in electrokinetics is proposed due to its capacity to mobilize Cd, Co, Cu and Zn. Low mobilization was observed for Cr and Pb;
- The increase of the treatment time and the concentration of citric acid up to 0.5 M resulted in the effective removal of Cd, Co, Cu and Zn from aged contaminated soil. The removal ratios of Cd, Co, Cu and Zn were over 70–80%. It is expected to achieve complete remediation of these metals, increasing the treatment time;
- The remediation of soil contaminated with Cr and Pb requires further research to identify the appropriate experimental conditions that lead to significant mobilization and removal;
- The metal soil extraction tests suggest that oxalic acid could be an effective extractant for Cr and acetic acid for Pb. These two organic acids could be tested in combination with citrate for the simultaneous removal of the six metals. Alternatively, sequential electrokinetic treatment with citric acid, oxalic acid and acetic acid may result in the removal of the six metals.

Author Contributions: Conceptualization, C.C.; methodology, C.C.; validation, C.C., S.G. and A.C.; formal analysis, C.C., S.G. and A.C.; investigation, C.C., S.G. and A.C.; resources, C.C. and S.G.; data curation, C.C., S.G. and A.C.; writing—original draft preparation, C.C.; writing—review and editing, C.C., S.G. and A.C.; visualization, C.C., S.G., and A.C. All authors have read and agreed to the published version of the manuscript.

Funding: This research received no external funding.

Data Availability Statement: Data are contained within the article.

Conflicts of Interest: The authors declare no conflict of interest.

References

1. Arao, T.; Ishikawa, S.; Murakami, M.; Abe, K.; Maejima, Y.; Makino, T. Heavy metal contamination of agricultural soil and countermeasures in Japan. *Paddy Water Environ.* **2010**, *8*, 247–257. [CrossRef]
2. Van Straalen, N.M. Assessment of soil contamination—A functional perspective. *Biogeochem.* **2002**, *13*, 41–52. [CrossRef] [PubMed]
3. Krishna, A.K.; Govil, P.K. Soil Contamination Due to Heavy Metals from an Industrial Area of Surat, Gujarat, Western India. *Environ. Monit. Assess.* **2006**, *124*, 263–275. [CrossRef] [PubMed]
4. Liu, L.; Li, W.; Song, W.; Guo, M. Remediation techniques for heavy metal-contaminated soils: Principles and ap-plicability. *Sci. Total Environ.* **2018**, *633*, 206–219. [CrossRef] [PubMed]
5. Hou, D.; O'Connor, D.; Igalavithana, A.D.; Alessi, D.S.; Luo, J.; Tsang, D.C.W.; Sparks, D.L.; Yamauchi, Y.; Rinklebe, J.; Ok, Y.S. Metal contamination and bioremediation of agricultural soils for food safety and sustainability. *Nat. Rev. Earth Environ.* **2020**, *1*, 366–381. [CrossRef]
6. Khalid, S.; Shahid, M.; Niazi, N.K.; Murtaza, B.; Bibi, I.; Dumat, C. A comparison of technologies for remediation of heavy metal contaminated soils. *J. Geochem. Explor.* **2017**, *182*, 247–268. [CrossRef]
7. Si, Y.; Lao, J.; Zhang, X.; Liu, Y.; Cai, S.; Gonzalez-Vila, A.; Li, K.; Huang, Y.; Yuan, Y.; Caucheteur, C.; et al. Electrochemical Plasmonic Fiber-optic Sensors for Ultra-Sensitive Heavy Metal Detection. *J. Light. Technol.* **2019**, *37*, 3495–3502. [CrossRef]
8. Hwang, J.-H.; Wang, X.; Zhao, D.; Rex, M.M.; Cho, H.J.; Lee, W.H. A novel nanoporous bismuth electrode sensor for in situ heavy metal detection. *Electrochim. Acta* **2019**, *298*, 440–448. [CrossRef]

9. Reddy, K.R.; Cameselle, C. *Electrochemical Remediation Technologies for Polluted Soils, Sediments and Groundwa-Ter*; John Wiley & Sons: Hoboken, NJ, USA, 2009.
10. Karaca, O.; Cameselle, C.; Bozcu, M. Opportunities of electrokinetics for the remediation of mining sites in Biga peninsula, Turkey. *Chemosphere* **2019**, *227*, 606–613. [CrossRef] [PubMed]
11. Ricart, M.T.; Cameselle, C.; Lucas, T.; Lema, J.M. Manganese Removal from Spiked Kaolinitic Soil and Sludge by Electromigration. *Sep. Sci. Technol.* **1999**, *34*, 3227–3241. [CrossRef]
12. Cameselle, C.; Gouveia, S. Removal of Multiple Metallic Species from Sludge by Electromigration. *J. Hazard. Toxic Radioact. Waste* **2020**, *24*, 04019030. [CrossRef]
13. Ottosen, L.M.; Hansen, H.K.; Bech-Nielsen, G.; Villumsen, Å. Electrodialytic Remediation of an Arsenic and Copper Polluted Soil—Continuous Addition of Ammonia during the Process. *Environ. Technol.* **2000**, *21*, 1421–1428. [CrossRef]
14. Cameselle, C.; Pena, A. Enhanced electromigration and electro-osmosis for the remediation of an agricultural soil contaminated with multiple heavy metals. *Process. Saf. Environ. Prot.* **2016**, *104*, 209–217. [CrossRef]
15. Zhu, Y.; Fan, W.; Zhou, T.; Li, X. Removal of chelated heavy metals from aqueous solution: A review of current methods and mechanisms. *Sci. Total. Environ.* **2019**, *678*, 253–266. [CrossRef] [PubMed]
16. ASTM. *ASTM D1293 Standard Test Methods for pH of Water. ASTM International*; ASTM International: West Conshohocken, PA, USA, 2018. [CrossRef]
17. ASTM. *ASTM D854 Standard Test Methods for Specific Gravity of Soil Solids by Water Pycnometer*; ASTM International: West Conshohocken, PA, USA, 2014. [CrossRef]
18. ASTM. *ASTM D2980 Standard Test Method for Saturated Density, Moisture-Holding Capacity, and Porosity of Saturated Peat Materials*; ASTM International: West Conshohocken, PA, USA, 2017. [CrossRef]
19. ASTM. *ASTM D2216 Standard Test Methods for Laboratory Determination of Water (Moisture) Content of Soil and Rock by Mass*; ASTM International: West Conshohocken, PA, USA, 2019. [CrossRef]
20. ASTM. *ASTM D2974 Standard Test Methods for Determining the Water (Moisture) Content, Ash Content, and Organic Material of Peat and Other Organic Soils*; ASTM International: West Conshohocken, PA, USA, 2020. [CrossRef]
21. ASTM. *ASTM D422 Standard Test Method for Particle-Size Analysis of Soils*; ASTM International: West Conshohocken, PA, USA, 2007. [CrossRef]
22. USEPA. Acid Digestion of Aqueous Samples and Extracts for Total Metals for Analysis by FLAA or ICP Spectroscopy. In *Test Methods for Evaluating Solid Waste, Physical/Chemical Methods*; USEPA Method 3010A; EPA SW-846; United States Environmental Protection Agency, Office of Solid Waste: Washington, DC, USA, 1992.
23. USEPA. Acid Digestion of Sediments, Sludges, and Soils. In *Test Methods for Eval-uating Solid Waste, Physical/Chemical Methods*; EPA SW-846; US EPA Method 3050B Revision 2; United States Environmental Protection Agency, Office of Solid Waste: Washington, DC, USA, 1996.
24. Morel, F.M.; Hering, J.G. *Principles and Applications of Aquatic Chemistry*; John Wiley & Sons: Hoboken, NJ, USA, 1993.
25. Cameselle, C.; Gouveia, S.; Urréjola, S. Benefits of phytoremediation amended with DC electric field. Application to soils contaminated with heavy metals. *Chemosphere* **2019**, *229*, 481–488. [CrossRef] [PubMed]
26. Clausen, C. Improving the two-step remediation process for CCA-treated wood: Part I. Evaluating oxalic acid extraction. *Waste Manag.* **2004**, *24*, 401–405. [CrossRef] [PubMed]
27. Renella, G.; Landi, L.; Nannipieri, P. Degradation of low molecular weight organic acids complexed with heavy metals in soil. *Geoderma* **2004**, *122*, 311–315. [CrossRef]
28. Figueroa, A.; Cameselle, C.; Gouveia, S.; Hansen, H.K. Electrokinetic treatment of an agricultural soil contaminated with heavy metals. *J. Environ. Sci. Health Part A* **2016**, *51*, 691–700. [CrossRef] [PubMed]
29. Nörtemann, B. Biodegradation of EDTA. *Appl. Microbiol. Biotechnol.* **1999**, *51*, 751–759. [CrossRef] [PubMed]
30. Abbay, G.; Gilbert, T. Chromium(III)-citrate complexes: A study using ion exchange and isotachophoresis. *Polyhedron* **1986**, *5*, 1839–1844. [CrossRef]

Article

Testing of Natural Sorbents for the Assessment of Heavy Metal Ions' Adsorption

Vera Yurak [1,2,*], Rafail Apakashev [1], Alexey Dushin [1], Albert Usmanov [1], Maxim Lebzin [1] and Alexander Malyshev [1]

[1] Research Laboratory of Disturbed Lands' and Technogenic Objects' Reclamation, Ural State Mining University, 620144 Yekaterinburg, Russia; parknedra@yandex.ru (R.A.); rector@m.ursmu.ru (A.D.); albert3179@mail.ru (A.U.); az_ma@mail.ru (M.L.); malyshev.k1b@gmail.com (A.M.)
[2] Center for Nature Management and Geoecology, Institute of Economics, The Ural Branch of Russian Academy of Sciences, 620014 Yekaterinburg, Russia
* Correspondence: vera_yurak@mail.ru

Abstract: Nowadays, the sorption-oriented approach is on the agenda in the remediation practices of lands contaminated with heavy metals. The current growing quantity of research accounts for different sorbents. However, there is still a lack of studies utilizing the economic criteria. Therefore, to ensure a wide application of opportunities, one of the necessary requirements is their economic efficiency in use. By utilizing these criteria, this manuscript researches the generally accepted natural sorbents for the assessment of heavy metal ions' adsorption, such as peat, diatomite, vermiculite and their mixtures in different proportions and physical shapes. The methodological base of the study consists of the volumetric (titrimetric) method, X-ray fluorescence spectrometry and atomic absorption spectrometry. Experimental tests show a certain decline in the efficiency of heavy metal ions' adsorption from aqueous salt solutes as follows: granular peat–diatomite > large-fraction vermiculite > medium-fraction vermiculite > non-granular peat–diatomite > diatomite.

Keywords: natural sorbents; ameliorants; disturbed lands; heavy metals; adsorption; recultivation; environmental remediation; revitalization; renaturation; restoration; rehabilitation; reclamation

1. Introduction

Modern intensive industrial production is accompanied by the formation of various man-made wastes. These wastes need to be neutralized and disposed. Wastes containing compounds of toxic chemical elements—heavy metals—are particularly dangerous. Heavy metals are pollutants which can be accumulated in the environment for a long time. Heavy metals are stable to biological and chemical degradation [1–5].

Nowadays, high research interest in the detection and sorption of heavy metal ions has been observed, including the development of novel sensors [6,7] because at relatively low concentrations, heavy metals are dangerous for soil, plants, living organisms and human health [8,9]. That is why technologies aimed at limiting the mobility of heavy metals are increasingly used [10]. Special functional substances—natural and synthetic sorbents—reduce the mobility of toxic elements by immobilizing heavy metal ions. Sorbents fix the mobile forms of heavy metals and significantly reduce their invasion into the biomass of plants, animals, fish and the human body.

Currently, the sorption-oriented approach is widely used. It acts as a tool for restoring disturbed land after any anthropogenic activity. This is due to the fact that natural soil remediation from heavy metals is very slow. For instance, the half-lives of lead, copper and cadmium are equal to 740–5900, 310–1500 and 13–110 years, respectively [11].

However, some of the traditional methods for the deactivation of heavy metals are extremely expensive. In this regard, they are used to isolate certain small contaminated

areas [12]. Therefore, researchers from different countries are currently studying the effectiveness of different sorbents. The core aim is the decontamination of heavy metals in mining and other industries. Natural sorbents are interesting due to their low cost, efficiency and availability of large reserves. They have the functional ability to act not only as sorbents, but also as ameliorants [13–15].

Organic raw materials of plant origin, such as peat, coal, sapropels, diatomite, vermiculite, industrial wood residue and agricultural waste present the greatest practical interest for the development of sorbent-ameliorants. The issue of bioremediation, using microorganisms in situ to decompose pollutants, is also relevant. In recent years, this technology has attracted significant attention from scientists and biotechnologists focused on practical research [16,17].

As a natural sorbent, diatomite modified with polyhydroxyl-aluminum was researched, particularly its increasing adsorption capacity to Pb^{2+} and Cd^{2+}. The results showed that diatomite modified with polyhydroxyl-aluminum significantly improved the adsorption of Pb^{2+} and Cd^{2+} by 23.79% and 27.36%, respectively [18].

There was research about peat's (from the Vale do Ribeira in Brazil) adsorption capacity of lead (Pb) and cadmium (Cd). The obtained results demonstrated that peat could improve the characteristics of soil contaminated with heavy metals. The researchers concluded that soil–peat mixtures were capable of minimizing the potentially toxic metal contamination. In addition, these are affordable and cheap materials [19].

Numerous studies confirm that adsorbents such as clay, coal, peat moss, zeolite and chitosan are affordable and effective for the remediation of land contaminated with heavy metals. However, despite the significant adsorption characteristics and low cost, there is a lack of a comprehensive studies of the sorption abilities of natural adsorbents [20,21], except for a few studies conducted a long time ago [22–25]. This article is an attempt to compare the sorption abilities of common natural substances, including peat, diatomite, vermiculite and their mixtures, in different proportions and physical shapes.

2. Materials and Methods

2.1. Materials

The cost of sorbents is one of the necessary requirements for sorbents to ensure the possibility of wide application. In this regard, it is rational to use primarily inexpensive materials as sorbents [26]. Based on these economic criteria, this paper researches the generally accepted natural sorbents for the assessment of heavy metal ions' adsorption, such as peat, diatomite, vermiculite and their mixtures in different proportions and physical shapes.

The following materials have been employed in the study (Table 1).

Table 1. Characteristics of the natural sorbents discussed in the study.

Sorbent Name	Ratio (%)	pH	Moisture (%)	Ash Content (%)	Apparent Density (kg/m^3)	Specific Surface Area of Grains (m^2/kg)	Temperature (K)
1. Granular peat-diatomite	50%	6.5–7.5	15	Less than 5	215	1.94	295
2. Non-granular peat-diatomite	50%	6.5–7.5	25	Less than 5	272	0.05	295
3. Diatomite	100%	8.5–9.5	2	Less than 1	465	0.15	295
4. Large-fraction vermiculite	100%	5.5–6.0	8	Less than 1	134	3.24	295
5. Medium-fraction vermiculite	100%	5.5–6.0	8	Less than 1	145	2.17	295

The first material was neutralized high-moor fractionated peat (fraction 0–10), with a peat moisture content from 50% to 60%, pH in the range of 5.5–6.0 and ash content of less than 5%. The main inorganic peat compounds were nitrogen (up to 1.5%), phosphorus, potassium and calcium (in total) up to 0.6%. The content of humic substances was 7.4–7.9%.

Peat was mixed with diatomite in the mass ratio of 2/1. Diatomite is a soft, lightweight, thin-pored massive material consisting of the mass of the smallest (0.01–0.04 mm) opal shells of bacillariophytes. Their density is usually nearly 0.5–0.7 g/cm^3, and its porosity reaches 70–75% [27].

The apparent density of the peat and diatomite mixture was 272 kg/m³. Parts of this mixture were formed in granules in the screw pelletizer with further drying in the drum-type drier at a temperature of 80 °C. The heat treatment of the pellets was completed when the humidity reached 25%. The apparent density of the absorbent grains was 242 kg/m³, taking into account that the specific surface area of the grains was 1.94 m²/kg.

The second material was medium-fraction vermiculite (fraction 1–4 mm) with an apparent density of 145 kg/m³, as well as large-fraction vermiculite (fraction 5–10 mm) with an apparent density of 134 kg/m³ and a specific surface area of 3.24 m²/kg.

Vermiculite is a mineral from the hydromicas group which turns into a flowing laminal material during heat treatment [28]. Vermiculite improves the soil structure, absorbs the excess moisture, loosens the soil and increases the soil breathability.

The third material was model toxicants, which are the individual solutions of salts of heavy metals and arsenic related to them, including $CuSO_4 \cdot 5H_2O$, $Pb(NO_3)_2$, $Cd(NO_3)_2 \cdot 4H_2O$, $Cr_2(SO_4)_3 \cdot 6H_2O$, $K_2Cr_2O_7$ and $NaAsO_2$. To prepare the solution, a sample of salt was dissolved in a measuring container. Solutions with lower concentrations were prepared from the resulting initial solution by dilution. As a result, solutions of salts with the following titers (content) of heavy metals were procured: $T(Cr^{+3}) = 1.0$ mg/L, $T(Cr^{+6}) = 0.5$ mg/L, $T(Pb^{+2}) = 1400$ mg/L, $T(Cu^{+2}) = 635$ mg/L, $T(Cd^{2+}) = 10000$ mg/L and $T(As^{3+}) = 5.0$ mg/L. The ionic strengths of the prepared solutions were as follows: $\mu(CuSO_4) = 0.04$ mol/L, $\mu(Pb(NO_3)_2) = 0.02$ mol/L, $\mu(Cd(NO_3)_2) = 0.267$ mol/L, $\mu(Cr_2(SO_4)_3) = 0.0001$ mol/L and $\mu(NaAsO_2) = 0.0001$ mol/L. The calculated value of the ionic strength of the prepared sodium bichromate solution was close to zero. Standard (calibration) salt solutions as measuring instruments were prepared from the initial solutions by dilution. During the tests of the sorption properties of the materials, the mixing of different salts did not occur. Therefore, the formation of any precipitation was not observed during sorption in the heavy metal solutions.

2.2. Methods

The sorption properties of the sorbents were studied in static conditions at room temperature (T = 295 K). The sorbent subsamples were placed in glass flasks. The subsamples were taken with a weighing accuracy of ±0.01 g. A fixed volume of a heavy metal salt solute with the initial concentration was added to the flasks at pH = 4.5. The solutions' system of a sorbent and heavy metal salt solute obtained from flasks was filtered through a medium-density paper filter after a certain contact time of the components. To determine the concentrations of heavy metal ions in the solutions before and after adsorption, chemical methods and instrumental methods of quantitative analysis, such as the volumetric and titrimetric methods, X-ray fluorescence spectrometry and atomic absorption spectrometry, were used.

2.3. Volumetric and Titrimetric Methods of Quantitative Analysis

This method in the variant of complexometric titration was used to determine the quantitative content of Cd^{2+} ions with the indicator—eriochrome black T—in an ammonia buffer solution before the transition of the initial red color to blue. Trilon B was used as the standard solution. A blank (control) experiment was performed to determine the correction during titration. The blank experiment consisted of repeating the chemical analysis procedure under similar conditions, but with the previous addition of distilled water to the analyzed sorbent instead of the heavy metal salt solute. A relative determination error of the quantitative content of heavy metal ions by this titrimetric method was 0.3%.

2.4. X-ray Fluorescence Spectrometry

The X-ray spectrometry method was used to determine the quantitative content of Pb^{2+}, As^{3+}, Cr^{3+} and Cr^{6+} ions. The analysis was performed by using an X-ray fluorescence crystal diffraction scanning spectrometer called a SPECTROSCAN MAX G. Standard solutions of the corresponding salts were used as calibration samples during the analysis. Standard solutions were prepared according to the exact subsamples of the starting substance (±0.0001 g). The error of the X-ray fluorescence analysis, according to the passport of the device, varied in the range of 0.2–3%.

2.5. Atomic Absorption Spectrometry

This method of analysis was used to determine the quantitative content of Pb^{2+} and Cu^{2+} ions. The analysis was performed by using a Spectr AA-240 FS spectrometer (Varian Optical Spectroscopy Instruments, Varian Inc., Australia; it has been sourced from Hoogeveen, The Netherlands). Standard samples with a known content of heavy metal ions were used as calibration samples.

The method of atomic absorption spectrometry was characterized by the relative determination error of at least ±2%.

3. The Results of the Experimental Studies and Discussion

The results of testing the efficiency of the sorbents in relation to the heavy metal ions' adsorption are shown in Figure 1. The information in Figure 1 allowed for the evaluation of the efficiency of the studied sorbents in relation to the heavy metal ions' adsorption, which was shown during 70 h of the experiment. The maximum and close to the maximum possible heavy metal ion extraction from the solutions by sorbents of peat-diatomite and vermiculite of various fractions (medium and large) were noted. The granular sorbent peat–diatomite showed the greatest efficiency with respect to the ion binding of Pb^{2+} and Cu^{2+}. In this case, the degree and quantity of extraction of these ions from aqueous solutes of salts were 100.0 and 98.0%. The granular sorbent peat–diatomite was less effective in relation to Cr^{6+} ions (in the composition of $Cr_2O_7^{2-}$ ions), As^{3+} ions (AsO^{2-}) and Cd^{2+} ions. The degree of extraction of the listed ions by this sorbent was in a range from 60% to 80%. A fairly lower sorption efficiency of the granular sorbent peat–diatomite occurred with respect to the Cr^{3+} ions.

The non-granular sorbent peat–diatomite demonstrated nearly the same sorption efficiency as the granular sorbent peat–diatomite for the Pb^{2+}, Cu^{2+} and As^{3+} ions. This sorbent exhibited approximately half the capacity in relation to the chromium ions in a oxidation state +3, +6. For this sorbent, an anomalous decrease in the sorption capacity was observed in relation to the cadmium ions. The decrease occurred to a value which was comparable to the determination error of the quantitative content of the metal ions in the solution.

The high efficiency of adsorption in relation to the heavy metal ions in the experiments were noted for large- and medium-fraction vermiculite. Regardless of the fractional composition, vermiculite almost completely bound the ions of Cd^{2+}, Cu^{2+} and Pb^{2+}. At the same time, there was a five-fold difference in the sorption of As^{3+} ions. Large-fraction vermiculite sorbed almost all of the arsenic in the solution (medium-fraction vermiculite = only 18.5%). The probable reason for this discrepancy may be the use in the study of vermiculite from different manufacturers. They use various technological processing for natural source raw materials. Meanwhile, vermiculite of both fractions practically did not show the sorption properties in relation to the Cr^{6+} ions. The relevant experimental value of the adsorption was close to the value of the determination error.

A relatively lower efficiency in the binding of heavy metal ions in the research was identified for diatomite. This material, when used as a sorbent, adsorbed 88.9% of Tb^{2+} ions and 64.0% of Cu^{2+} ions from the solution. These were the best results for this sorbent. Diatomite virtually did not bind cadmium. However, diatomite was 2.5 times higher than the medium-fraction vermiculite in terms of As^{3+} ion adsorption.

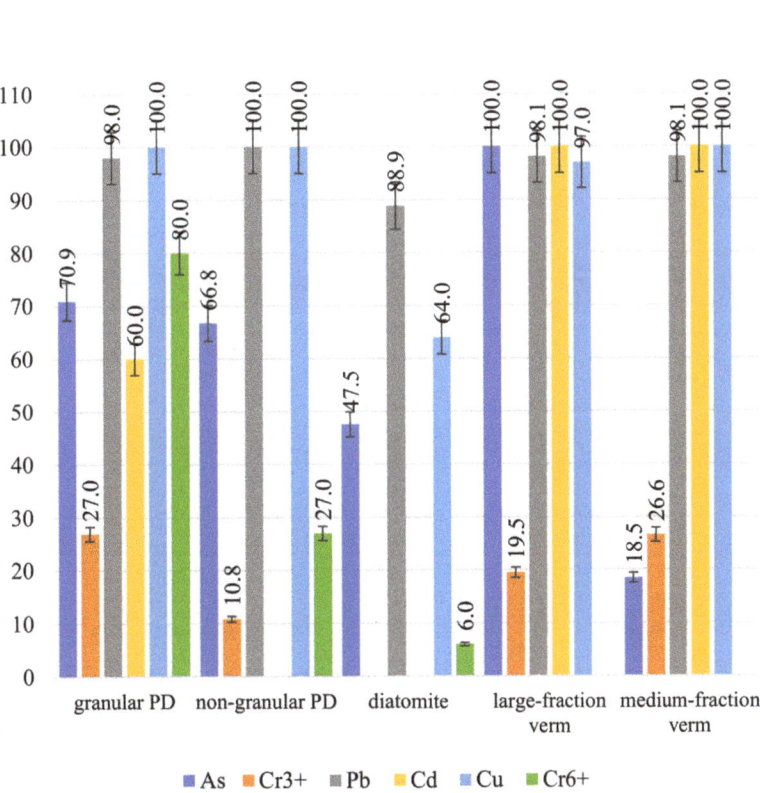

Figure 1. Efficiency of the sorbents (A) in relation to the grade of heavy metal ion extraction from the model solutions.

Figure 2 shows a general diagram for the studied sorbents and heavy metal salt solutions. Figure 2 demonstrates the total value of the adsorption of heavy metal ions. The granular sorbent peat–diatomite had the highest total efficiency in the sorption of the studied heavy metal ions. This was established based on the results of the sorbents' testing. Large-fraction vermiculite was almost as good as the granular sorbent peat–diatomite in terms of the indicator under consideration. The lowest efficiency was shown by the sorbent consisting of diatomite, which did not contain peat additives. Medium-fraction vermiculite and non-granular sorbent peat–diatomite took the intermediate position in terms of sorption efficiency.

The difference in the sorption efficiency of the studied sorbents was linked to their chemical compositions and the structure of the adsorbing surface. To fix the structure of the sorbents, an Altami MET 1 digital microscope and an SNE4500M scanning electron microscope were used. The sorption properties of peat were due to the highly developed surface (Figure 3) and the presence in its composition of various functional groups, such as amine, amide, alcohol, aldehyde, carboxyl, ketone, phenolic, kinone and peptide groups. The adsorptive properties of peat were due to the presence of lignin, humic and fulvic acids [29,30].

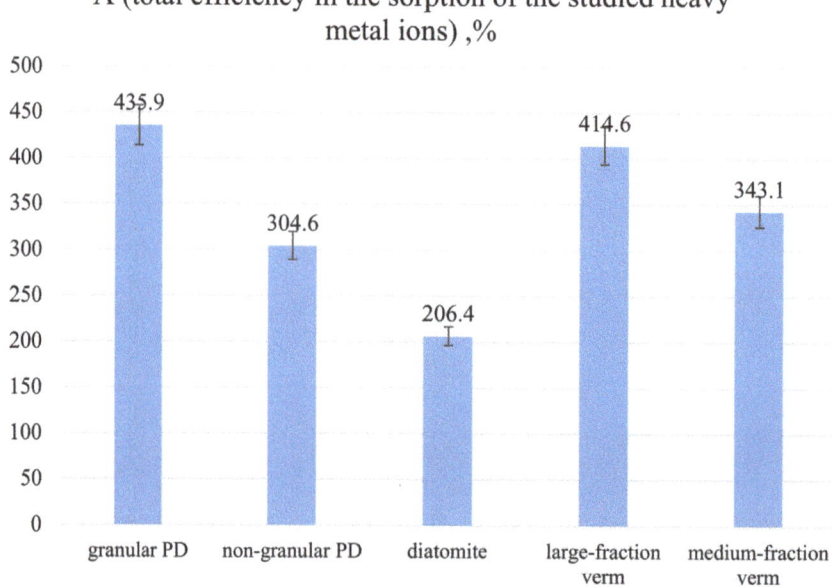

Figure 2. The total value of the degree of extraction of heavy metal ions (A) by various sorbents.

Figure 3. The surface of the peat–diatomite granular sorbent (Altami MET 1 optical microscope (Altami Ltd, Russia; it has been sourced from St. Petersburg, Russia); magnification ×40).

It should be noted that peat takes up a special place among the effective natural sorbents of heavy metals. Peat-based sorbents are good stabilizers of heavy metal ions. These sorbents have a higher sorption capacity than other single-component sorbents [31].

On the one hand, diatomite does not contain a larger number of different functional groups than peat. On the other hand, diatomite has a high porosity (Figure 4) and corresponding adsorption properties.

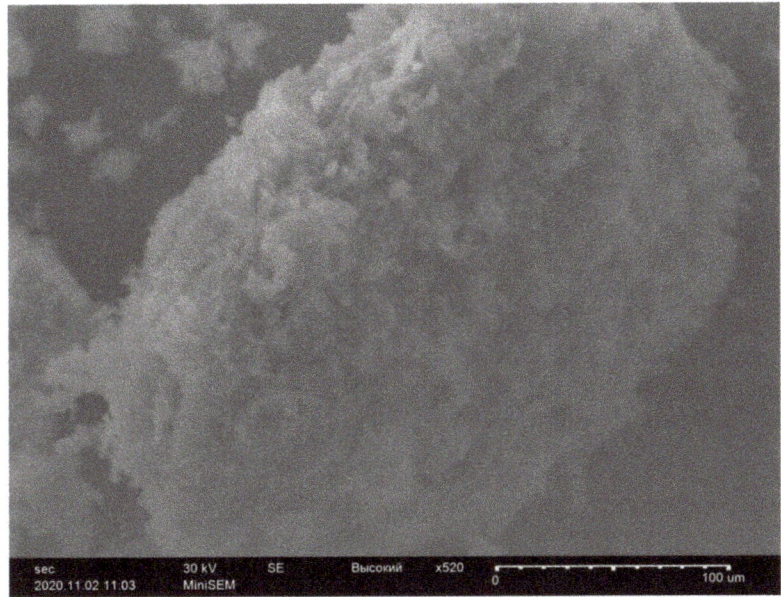

Figure 4. The surface of diatomite (SNE 4500M microscope).

Vermiculite is a mineral from the hydromica group. During heat treatment, vermiculite swells and multiplies in space, which also increases the area of the adsorbing surface (Figure 5). Exfoliated vermiculite has a variety of applications, including being a sorbent for gaseous and liquid industrial waste.

The study of the vermiculite's efficiency of heavy metal adsorption in comparison with other sorbents was also shown in [32], with the exception of peat–diatomite. In this research [32], Koptsik G.N. and Zakharenko A.I. used several types of ameliorants: vermiculite, zeolite, apatite superphosphate and lime. The residual copper content in the soils after the use of vermiculite on average did not exceed 9% of the initial concentration [32]. Their results are comparable to our study.

The use of peat and diatomite separately from each other was researched in [33,34]. In [34], filter modules were used to study the processes of sorption and desorption of emission components by materials of a vermiculite–sungulite composition, obtained through the enrichment of phlogopite mining waste. The study [34] concluded that pure peat had the greatest sorption capacity.

The study in [33] used diatomite rocks. The equilibrium state for the removal of heavy metals by diatomite clay was obtained by adding a constant mass of the absorbent (0.5, 1.0 and 2.0 g) to 100 mL of the initial concentrations of the prepared aqueous solution of heavy metals. The exposure time was 4 h. After that, the samples were filtered by using filter paper. The concentrations of various heavy metal ions were measured by using the ICP-MS method. It was concluded that the use of Egyptian diatomite demonstrated a significant efficiency of heavy metal adsorption [33,34].

The results of this study show a significant increase in the adsorption efficiency when two sorbents (peat–granular diatomite) are mixed in equal proportions (Figure 2). It is remarkable that the granular physical shape increases the heavy metal adsorption. Testing other proportions of different natural sorbents with various recipes of preparation, including different physical shapes, is the direction of future studies. Due to the research results, a patent has been obtained: "Peat ameliorant for the remediation of lands contaminated with heavy metals" RU 2 745 456, dated 9 March 2020.

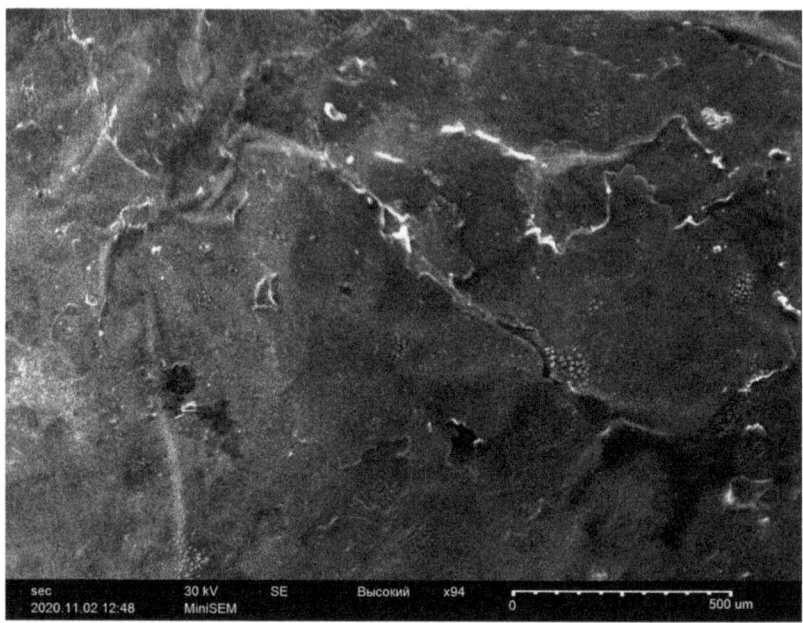

Figure 5. The surface of vermiculite (SNE 4500M microscope).

4. Conclusions

The testing results demonstrate a certain sequence in terms of the gradual decline of the efficiency of heavy metal ion adsorption from aqueous salt solutes (peat–granular diatomite > large-fraction vermiculite > medium-fraction vermiculite > peat–non-granular diatomite > diatomite).

The high efficiency of peat-containing sorbents in relation to the adsorption of heavy metals and the widespread use of peat as a soil ameliorant can serve as the basis for the promising use of peat in the development of compositions of affordable complex ameliorants, stabilizers of heavy metals for the remediation of disturbed lands.

Author Contributions: Conceptualization, data curation and project administration have been done by V.Y. All authors—V.Y., R.A., A.U., M.L., A.M. and A.D.—took part in the investigation. While supervision was conducted by R.A. and V.Y., the validation was made by M.L., A.U. and A.M. The original draft was written by R.A., A.U. and M.L. Review and editing were done by V.Y. and A.D. All authors have read and agreed to the published version of the manuscript.

Funding: This research was funded by the Ministry of Science and Higher Education of Russia in accordance with the state assignment for Ural State Mining University no. 075-03-2021-303 from 29 December 2020.

Institutional Review Board Statement: Not applicable.

Informed Consent Statement: Not applicable.

Data Availability Statement: Data is contained within the articles, open sources that mentioned in the reference list, and made up by authors during the experiments.

Conflicts of Interest: The authors declare no conflict of interest.

References

1. Seleznev, A.A.; Klimshin, A.V. Heavy metals in soils on the territory of Yekaterinburg. *Izvestija URSMU* **2020**, *1*, 96–104. [CrossRef]
2. Semenov, A.I.; Koksharov, A.V.; Pogodin, Y.I. The content of heavy metals in the soil of Chelyabinsk. *Meditsina Truda Ekol. Cheloveka* **2015**, *3*, 184–191.
3. Pisareva, A.V.; Belopukhov, S.L.; Savich, V.I.; Stepanova, L.P.; Gukalov, V.V.; Yakovleva, E.V.; Shaikhiev, I.G. Migration of heavy metals from the source of pollution depending on the relationships in the landscape. *Bull. Technol. Univ.* **2017**, *20*, 160–163.
4. Kheir, R.B.; Greve, M.; Greve, M.; Peng, Y.; Shomar, B. A Comparative GIS tree-pollution analysis between arsenic, chromium, mercury, and uranium contents in soils of urban and industrial regions in Qatar. *Euro Mediterranean J. Environ. Integr.* **2019**, *4*, 10. [CrossRef]
5. Mikkonen, H.G.; Dasika, R.; Drake, J.A.; Wallis, C.J.; Clarke, B.O.; Reichman, S.M. Evaluation of Environmental and Anthropogenic Influences on Ambient Background Metal and Metalloid Concentrations in Soil. *Sci. Total Environ.* **2018**, *624*, 599–610. [CrossRef]
6. Si, Y.; Lao, J.; Zhang, X.; Liu, Y.; Cai, S.; Gonzalez-Vila, A.; Li, K.; Huang, Y.; Yuan, Y.; Caucheteur, C.; et al. Electrochemical Plasmonic Fiber-optic Sensors for Ultra-Sensitive Heavy Metal Detection. *J. Light. Technol.* **2019**, *37*, 3495–3502. [CrossRef]
7. Roxby, D.N.; Rivy, H.; Gong, C.; Gong, X.; Yuan, Z.; Chang, G.-E.; Chen, Y.-C. Microalgae living sensor for metal ion detection with nanocavity-enhanced photoelectrochemistry. *Biosens. Bioelectron.* **2020**, *165*, 112420. [CrossRef]
8. Wuana, R.A.; Okieimen, F.E. Heavy Metals in Contaminated Soils: A Review of Sources, Chemistry, Risks and Best Available Strategies for Remediation. *ISRN Ecol.* **2011**, *2011*, 1–20. [CrossRef]
9. Dutta, S.; Mitra, M.; Agarwal, P.; Mahapatra, K.; De, S.; Sett, U.; Roy, S. Oxidative and genotoxic damages in plants in response to heavy metal stress and maintenance of genome stability. *Plant Signal. Behav.* **2018**, *13*, 1–49. [CrossRef] [PubMed]
10. Koptsik, G.N. Modern approaches to remediation of heavy metal polluted soils: A review. *Eurasian Soil Sci.* **2014**, *47*, 707–722. [CrossRef]
11. Nikovskaya, G.N.; Gruzina, T.G.; Ulberg, Z.R.; Koval, L.A.; Ovcharenko, F.D. Novel Approaches to Bioremediation and Monitoring of Soils Contaminated by Heavy Metals and Radionuclides. In *Role of Interfaces in Environmental Protection*; Barany, S., Ed.; Springer: Dordrecht, The Netherlands, 2003; pp. 529–536.
12. Singh, A.T.; Prasad, S.M. Remediation of heavy metal contaminated ecosystem: An overview on technology advancement. *Int. J. Environ. Sci. Technol.* **2015**, *12*, 353–366. [CrossRef]
13. Muhammad, Y.A.; Nenohai, A.J.W.T.; Mufti, N.; Situmorang, R.; Taufiq, A. Adsorption Properties of Magnetic Sorbent Mn0.25Fe2.75O4@SiO2 for Mercury Removal. *Key Eng. Mater.* **2020**, *851*, 197–204. [CrossRef]
14. Novikova, A.; Nazarenko, O.B.; Vtorushina, A.N.; Zadorozhnaya, T.A. Characterization of Badinsk Zeolite and its Use for Removal of Phosphorus and Nitrogen Compounds from Wastewater. *Mater. Sci. Forum* **2019**, *970*, 7–16. [CrossRef]
15. Lebedeva, I.; Lonzinger, T.M.; Skotnikov, V.A. Study of Sorption of Manganese (II) Cations by Composite Sorbent. *Mater. Sci. Forum* **2020**, *989*, 85–90. [CrossRef]
16. Mishra, M.; Mohan, D. Bioremediation of Contaminated Soils: An Overview. In *Adaptive Soil Management: From Theory to Practices*; Rakshit, A., Abhilash, P.C., Singh, H.B., Ghosh, S., Eds.; Springer: Singapore, 2017; pp. 323–337.
17. Mani, D.; Kumar, C. Biotechnological advances in bioremediation of heavy metals contaminated ecosystems: An overview with special reference to phytoremediation. *Int. J. Environ. Sci. Technol.* **2014**, *11*, 843–872. [CrossRef]
18. Zhu, J.; Wang, P.; Lei, M.-J.; Zhang, W.-L. Polyhydroxyl-aluminum pillaring improved adsorption capacities of Pb2+ and Cd2+ onto diatomite. *J. Central South Univ.* **2014**, *21*, 2359–2365. [CrossRef]
19. Marques, J.P.; Rodrigues, V.G.S.; Raimondi, I.M.; Lima, J.Z. Increase in Pb and Cd Adsorption by the Application of Peat in a Tropical Soil. *Water Air Soil Pollut.* **2020**, *231*, 1–21. [CrossRef]
20. Gürel, A. Adsorption characteristics of heavy metals in soil zones developed on spilite. *Environ. Earth Sci.* **2006**, *51*, 333–340. [CrossRef]
21. Maharana, M.; Manna, M.; Sardar, M.; Sen, S. Heavy Metal Removal by Low-Cost Adsorbents. In *Green Adsorbents to Remove Metals, Dyes and Boron from Polluted Water*; Environmental Chemistry for a Sustainable World; Inamuddin Ahamed, M.I., Lichtfouse, E., Asiri, A.M., Eds.; Springer International Publishing: Cham, Switzerland, 2021; Volume 49, pp. 245–272.
22. García-Sánchez, A.; Alastuey, A.; Querol, X. Heavy metal adsorption by different minerals: Application to the remediation of polluted soils. *Sci. Total. Environ.* **1999**, *242*, 179–188. [CrossRef]
23. Babel, S. Low-cost adsorbents for heavy metals uptake from contaminated water: A review. *J. Hazard. Mater.* **2003**, *97*, 219–243. [CrossRef]
24. Ríos, C.; Williams, C.; Roberts, C. Removal of heavy metals from acid mine drainage (AMD) using coal fly ash, natural clinker and synthetic zeolites. *J. Hazard. Mater.* **2008**, *156*, 23–35. [CrossRef]
25. Joseph, L.; Jun, B.-M.; Flora, J.R.; Park, C.M.; Yoon, Y. Removal of heavy metals from water sources in the developing world using low-cost materials: A review. *Chemosphere* **2019**, *229*, 142–159. [CrossRef]
26. Varank, G.; Demir, A.; Bilgili, M.S.; Top, S.; Sekman, E.; Yazici, S.; Erkan, H.S. Equilibrium and kinetic studies on the removal of heavy metal ions with natural low-cost adsorbents. *Environ. Protect. Eng.* **2014**, *40*, 43–61. [CrossRef]
27. Ubaskina, Y.; Korosteleva, Y. Investigation of the possibility of practical use of diatomit for wastewater treatment. *Bull. Belgorod State Technol. Univ. Named V.G. Shukhov* **2017**, *2*, 92–96.

28. Kalashnik, A.V.; Ionov, S.G. Obtaining and physico-chemical properties of materials based on expanded vermiculites of various compositions. *IVUZKKT* **2018**, *61*, 76–82. [CrossRef]
29. Gavrilov, S.V.; Kanarskaya, Z.A. Adsorption properties of peat and peat-based products. *Bull. Kazan Technol. Univ.* **2015**, *2*, 422–427.
30. Kuznetsova, I.A.; Larionov, N.S. The resource basis of north-west russia representative oligotrophic bogs: Chemical composition and binding properties. *ACNR* **2018**, *7*, 165–170. [CrossRef]
31. Soukand, U.; Kängsepp, P.; Kakum, R.; Tenno, T.; Mathiasson, L.; Hogland, W. Selection of adsorbents for treatment of leachate: Batch studies of simultaneous adsorption of heavy metals. *J. Mater. Cycles Waste Manag.* **2010**, *12*, 57–65. [CrossRef]
32. Koptsik, G.N.; Zakharenko, A.I. Effect of different amendments on mobility and toxicity of nickel and copper in polluted soils. *Bull. Mosc. State Univ.* **2014**, *1*, 32–37. [CrossRef]
33. Elsayed, E.E. Natural diatomite as an effective adsorbent for heavy metals in water and wastewater treatment (a batch study). *Water Sci.* **2018**, *32*, 32–43. [CrossRef]
34. Mosendz, I.A.; Kremenetskaya, I.P.; Drogobuzhskaya, S.V.; Alekseeva, S.A. Sorption of heavy metals by the filtering containers with serpentine materials. *Vestnik MGTU* **2020**, *23*, 182–189. [CrossRef]

Article

The Potential Effectiveness of Biochar Application to Reduce Soil Cd Bioavailability and Encourage Oak Seedling Growth

Elnaz Amirahmadi [1], Seyed Mohammad Hojjati [1], Claudia Kammann [2,*], Mohammad Ghorbani [3] and Pourya Biparva [4]

1. Department of Forest Sciences and Engineering, Sari Agricultural Sciences and Natural Resources University, Sari 4818168984, Iran; amirahmadielnaz@gmail.com (E.A.); s_m_hodjati@yahoo.com (S.M.H.)
2. Department for Applied Ecology/Climate Change Research for Special Crops, Geisenheim University, 65366 Geisenheim, Germany
3. Department of Soil Science, Faculty of Agricultural Sciences, University of Guilan, Rasht 4199613776, Iran; mghorbani0007@gmail.com
4. Department of Basic Sciences, Sari Agricultural Sciences and Natural Resources University, Sari 4818168984, Iran; p.biparva@sanru.ac.ir
* Correspondence: claudia.kammann@hs-gm.de

Received: 1 April 2020; Accepted: 2 May 2020; Published: 14 May 2020

Abstract: Today, it is very important to protect plants in soils contaminated with metals. We investigated the behavior of cadmium during the establishment of oak seedlings (*Quercus castaneifolia* C.A. Mey.) under biochar influence. This study was conducted in pots with loamy soil. Cadmium was added to soil at 0, 10, 30, and 50 mg per kg of soil, indicated by Control, Cd10, Cd30 and Cd50. Biochar was produced at 500–550 °C from rice husk and added at 1, 3, and 5% (wt/wt) levels, indicated by B1, B3, B5, and mixed with soil at planting in three replications. Generally, increasing biochar rates had significant effects on seedling height, diameter, and biomass. This coincided with Cd immobilization in the contaminated soil which reflects a decrease in Cd concentrations in the plant bioavailability of Cd. The tolerance index increased significantly, by 40.9%, 56%, and 60.6% in B1, B3, and B5 with Cd50, respectively, compared to polluted soil. The percent of Cd removal efficiency for Cd50 was 21%, 47%, and 67% in B1, B2, and B5, respectively. Our study highlights that biochar can reduce Cd bioavailability and improve the growth of oak seedlings in contaminated soil.

Keywords: seedling biomass; heavy metals; soil amelioration

1. Introduction

Anthropogenic activities that lead to producing landfill leachates, vehicle emissions, and the use of sewage water irrigation or pesticides can accumulate several metals in soils [1]. Recently, the concern for metal pollution in soil has increased and is recognized as a serious environmental problem [2,3]. Heavy metals remain in the soil for a long time and are irreversible [1]. Cadmium (Cd) is a toxic element that is non-essential for plant growth [2]. Large concentrations of Cd not only reduce growth and yield but also negatively affect plant physiological activities and can even cause plant death in higher concentrations [3,4]. Cadmium is dangerous because it is highly mobile in soil–plant systems, and it is highly toxic to plants [4]. Cadmium causes reductions in photosynthesis and subsequently in plant growth, and it can ultimately result in plant death [5]. Several biosynthesis processes have been prevented by excess Cd which disrupts the photosynthetic system in the plant. Cadmium stress, furthermore, induces necrosis, alters stomatal movements, ion homeostasis, and hence limits the availability of water and nutrients while it also affects the activities of several key enzymes and respiration in plants [6].

Applying organic amendments of plant or animal origin has the advantage of improving soil C sequestration and is an eco-friendly method for reducing the toxicity of metals [7]. Biochar is a soil modifier that is produced through pyrolysis of feedstock, such as agricultural residues, sawdust, wood chips, etc., under controlled oxygen conditions and temperature ranging from 350 to 750 °C [8]. Mohamed et al. [9] reported that biochars can retain nutrients and organic/inorganic contaminants. Biochar can trap metals and reduce their toxic effects due to the fact of its high cation exchange capacity (CEC), high pH, and large effective surface area [10–13]. Biochar is generally rich in nutrients for the plant and when added, it increases them in the soil. [8]. Cui et al. [7] examined the effect of RHB on the soil properties of an acid sulfate soil and observed an increase in soil organic carbon (SOC), soil pH, CEC, phosphorus (P), potassium (K), and calcium (Ca) and a decrease in exchangeable Al and soluble Fe.

Biochar can alter soil properties in such a way that it reduces the mobility of inorganic elements such as Cd, subsequently increasing crop production but also nutrient retention in soil [14].

Quercus castaneifolia (Oak) is one of the major species with industrial valuable in the northern forests of Iran with a large distribution in the Caspian Hyrcanian mixed forests (36°16′17.9″ N 57°7′32.9″ E). Oak seeds and leaves are used in pharmaceuticals, dyeing (as one of the natural dyes), and in the leather industry [15]. Unfortunately, human degradation in nature, such as mining, has caused the toxic metal contamination of Hyrcanian forest soil. Hence, protecting the Hyrcanian forest, as one of the UNESCO's World Heritage Sites, with valuable native species is necessary. Therefore, the use of biochar as a soil amendment can be a suitable approach to improve soil quality [16] and, at the same time, improve soil C sequestration.

The present study was therefore aimed at investigating the behavior of cadmium during the establishment of oak seedlings under biochar influence and performance, i.e., to examine if the use of biochar can serve as a tool for reclamation programs in mining areas.

2. Materials and Methods

2.1. Plant and Soil Preparation

The seedlings were obtained from Savadkoh Forest Nursery, Mazandaran (36°16′17.9″ N 57°7′32.9″ E). Cadmium nitrate $Cd(NO_3)_2$ as Cd solution with deionized water was used to make contamination treatments [16]. Three concentrations of Cd, including 0, 10, 30, and 50 mg per kg of soil, indicated by Control, Cd10, Cd30 and Cd50, were used in the experiment [17]. Cadmium levels were combined with three levels of biochar including 1% (B1), 3% (B2), and 5% (B5) by weight. The experimental design is presented in Table 1. The pot experiment (resulting in 39 pots in total) was conducted at Sari Agricultural Science and Natural Resources University (SANRU), Mazandaran Province, Iran (36°34′46.0″ N 53°11′30.4″ E). Each seedling was planted separately in 3 kg plastic pots on 20 March 2017.

Table 1. Experimental design for treatments.

Treatments	Cd10	Cd30	Cd50
Polluted soil	Cd10	Cd30	Cd50
Control	-	-	-
B1	B1 + Cd10	B1 + Cd30	B1 + Cd50
B3	B3 + Cd10	B3 + Cd30	B3 + Cd50
B5	B5 + Cd10	B5 + Cd30	B5 + Cd50

The properties of the soil used in this study are presented in Table 2.

Table 2. Physicochemical characteristics (mean values ± SE) of the forest nursery soil before the start of the pot experiment.

Physicochemical Parameter	Amount
pH	6.67 ± 0.01
EC (dS/m)	819 ± 1.08
OC (%)	6.53 ± 0.23
N (%)	0.082 ± 0.003
P (mg kg^{-1})	29.17 ± 0.66
K (mg kg^{-1})	550.39 ± 6.73
CEC (cmol$^{(+)}$ kg^{-1})	7.97 ± 0.25
Saturation Fluid Moisture (%)	53.65 ± 0.48
FC (%)	32 ± 0.52
(% Sand:Silt:Clay)	50.1:31.8:18.1
Soil Texture	Loamy

EC: electrical conductivity, OC: organic matter, N: nitrogen, P: phosphorus, K: potassium, CEC: cation exchangeable capacity, FC: field capacity.

2.2. Climatic Conditions and Irrigation

Climatic conditions during the experimental period are shown in Figure 1. The Fifty-year average annual rainfall sum was 768 mm, and the mean annual temperature was 19.3 °C.

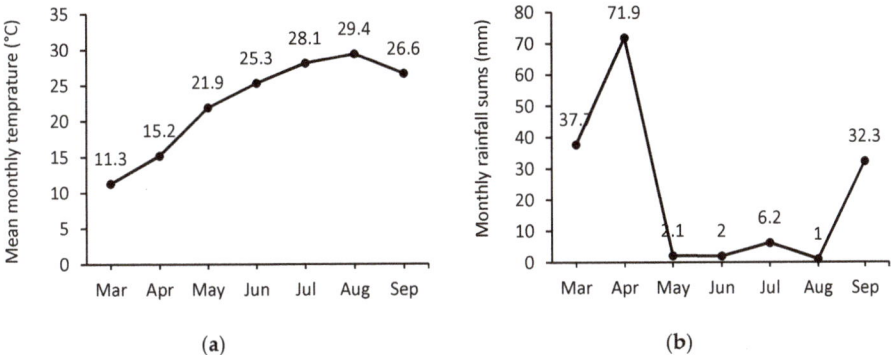

Figure 1. Climatic conditions during the experimental period: (a) mean monthly temperatures; (b) monthly rainfall sums.

We used tap water to irrigate the seedlings over the entire experiment. Irrigation was applied during the initial establishment of the experiment and, depending on the needs of the seedlings, two to three times a week where all tree seedlings received equal amounts of water. The amount of irrigation water was based on field capacity. Irrigation water for the whole period had neutral pH and concentrations of Cd (0.005 ppm).

2.3. Feedstock and Properties of Biochar

The biochar used in this study was produced from rice husk at 550 °C. [18,19]. The biochar was characterized by elemental analysis (VarioMax CHNO Analyzer). Electrical conductivity and pH were measured using a 1:20 (biochar:water) solution [20]. The detailed chemical and physical characterizations of rice husk biochar are given in Table 3. The amounts of 1%, 3%, and 5% by weight [1] of rice husk biochar were homogeneously mixed with the soil one day before planting the seedlings.

Table 3. Characteristics of rice husk biochar.

Indicators	Value	Unit	Characterizations	Value	Unite
pH	8.14	-	Oxygen	0.001	%
EC	359	dS m^{-1}	Phosphorous	412	mg kg^{-1}
H/C	0.36	Molar ratio	Sodium	76.1	mg kg^{-1}
C/N ratio	101.7	-	Potassium	595	mg kg^{-1}
CEC	18.28	cmol(+) kg^{-1}	Calcium	609	mg kg^{-1}
Carbon	68.03	%	Magnesium	163	mg kg^{-1}
Nitrogen	0.64	%	Iron	65	mg kg^{-1}
Hydrogen	25.12	%	Zinc	11.5	mg kg^{-1}

EC, electrical conductivity; CEC, cation exchangeable capacity; H/C, hydrogen/carbon ratio; C/N, carbon/nitrogen ratio.

2.4. Seedling Height, Growth, and Biomass Analysis

The height and diameter of the growing seedlings were measured once monthly, and the biomass of the seedlings was recorded after harvest at the end of the experiment (October 15, 2017). After the end of the experiment, the seedlings were washed with distilled water and then cut into leaves, stems, and roots to obtain a constant weight at 70 °C. In order to measure cadmium concentration in plant organs, a laboratory mixer (IKA Labortechnik M20) was used to powder the samples.

2.5. Selected Soil Properties and Heavy Metal Detection

The air-dried soil samples were analyzed for EC (by 1:5/soil:water suspension), pH (by 1:2.5/soil:water suspension), OC (by Walkley–Black method) [21], and CEC (by EDTA extraction method) [22]. The total nitrogen of the soil was measured by Kjeldahl and available phosphorus was measured by extraction with sodium bicarbonate method [23]. Measurement of soluble potassium in soil saturation extract was measured following the method of Helmke and Sparks [24]. The concentration of potassium in the extracts was measured using the Jenway PFP7 Flame Photometer.

Bioavailable cadmium was determined by the single extraction method [25]. To determine Cd concentrations in powdered plant material, 0.5 g of homogenized samples, 0.5 ml of 37% hydrochloric acid, 9.0 mL of 69% nitric acid, and 1 mL of 30% hydrogen peroxide was added into the digestion vessel in an open system. All chemical materials were purchased from Sigma-Aldrich Corporation. After the digestion program, the Cd concentration in samples was analyzed by an atomic absorption spectrophotometer (PinAAcle 900F).

The tolerance index (TI%) [26] and removal efficiency (RE%) [27] were calculated by the following formulas:

$$TI(\%) = \left(1 - \frac{\overline{X_f}}{\overline{X_c}}\right) \times 100 \quad (1)$$

where $\overline{X_f}$ is the mean biomass in polluted soil, and $\overline{X_c}$ is the mean biomass in the control.

$$RE(\%) = \left[1 - \frac{C}{C_0}\right] \times 100 \quad (2)$$

where C_0 and C are the initial and final extractable Cd concentrations in the soil, respectively.

2.6. Data Analysis

The experiment was based on one-factorial with three replications and the Student-Newman-Keuls (SNK)test was performed to test for significant differences among treatments ($p < 0.05$). The data's normal distribution test was evaluated using the Kolmogorov–Smirnov test and homogeneity of variance was checked using Levene's test. SPSS 23.0 was used to analyze the data.

3. Results

3.1. Seedling Growth

Polluted soil significantly decreased the diameter of seedlings and height compared to the control ($p < 0.05$) (Figure 2). There was no significant difference in the diameter (Figure 2a) and height (Figure 2b) seedlings among treatments containing biochar and the control, except for treatment B1. However, no significant difference was observed between Cd levels in biochar treatments. But, in general, the polluted soil (without biochar) had a significantly lower diameter and height of seedlings than the biochar treatments.

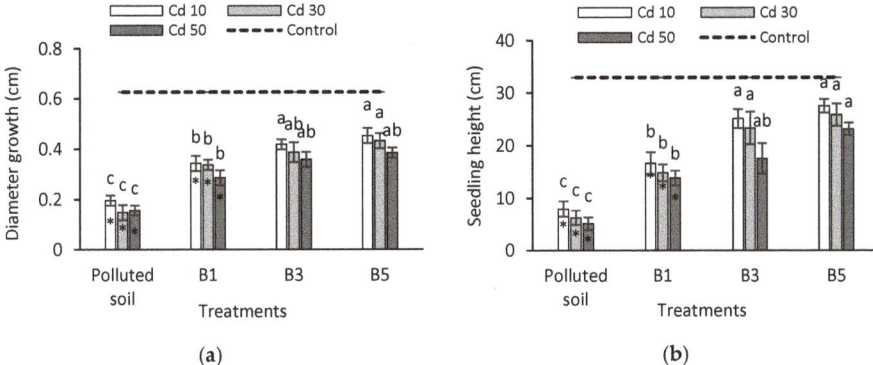

Figure 2. Effect of different levels of biochar on diameter growth and seedling height in oak seedlings at different levels of Cd contamination (means +/− standard deviation): diameter growth (**a**) and seedling height (**b**). The dashed line represents the control (no biochar, no Cd), and the term "Polluted soil" indicates pots with Cd but without biochar. Different letters represent significant differences between treatments. The star indicates a significant difference between treatments and control.

3.2. Seedling Biomass

Cadmium treatments significantly decreased the leaf dry biomass, stem dry biomass, root dry biomass, and total dry biomass of the oak seedlings (Figure 3), stronger so with increasing Cd amendment level ($p < 0.05$), while biochar amendments increased the biomasses of all plant organs compared to the same pollution levels without the biochar amendment. Again, an amendment of 5% of biochar always showed the best results at all contamination levels of Cd.

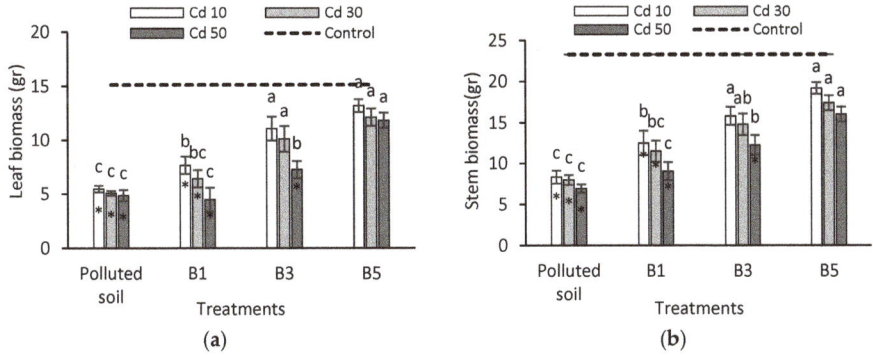

Figure 3. Cont.

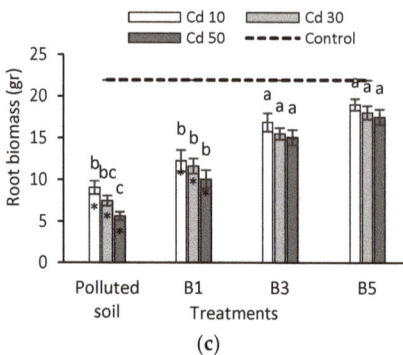
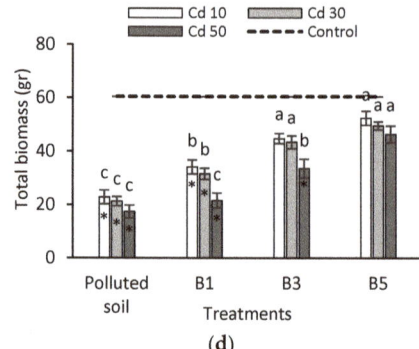

Figure 3. Effect of different levels of biochar on: (**a**) leaf biomass; (**b**) stem biomass; (**c**) root biomass; and (**d**) total biomass in oak seedlings at different levels of Cd contamination (bars represent means +/− one standard deviation). The dashed line represents the control (no biochar, no Cd), and the term "Polluted soil" indicates pots with the Cd but without biochar. Different letters represent significant differences between treatments. The star indicates a significant difference between treatments and control.

3.3. Chemical Soil Properties

The results showed that only the main effect of biochar on soil properties was significant ($p < 0.05$). In the case of soil pH, the application of 3% and 5% biochar showed significant increases than control (from 6.42 to 7.17 and 7.45, respectively) (Table 4). There was no significant difference in soil EC between biochar application rates and control (Table 4). B3 and B5 caused a significant increase ($p < 0.05$) in the OC content. Also, B3 and B5 significantly increased CEC ($p < 0.05$) by 9% and 20%, respectively (Table 4). In the case of N, P, and K, all the treatments with biochar had significantly higher values than the control ($p < 0.05$).

Table 4. Soil chemical properties (mean values ± SE) in presence of various percent of biochar.

Percent of Biochar	pH	EC (dS m^{-1})	OC (%)	CEC (cmol(+) kg^{-1})	N (mg kg^{-1})	P (mg kg^{-1})	K (mg kg^{-1})
0 (Control)	6.4 ± 0.02 [b]	0.31 ± 0.02 [a]	1.1 ± 0.02 [b]	8.11 ± 1.02 [c]	0.15 ± 0.01 [c]	12.1 ± 1.2 [c]	8.3 ± 1.2 [c]
1%	6.5 ± 0.01 [b]	0.27 ± 0.05 [a]	1.2 ± 0.04 [b]	12.31 ± 1.01 [b]	0.4 ± 0.03 [b]	21.1 ± 2.1 [b]	15.4 ± 1.1 [b]
3%	7.17 ± 0.03 [a]	0.26 ± 0.02 [a]	1.57 ± 0.01 [a]	13.35 ± 1.03 [ab]	0.5 ± 0.05 [b]	35.5 ± 3.3 [a]	29.5 ± 1.3 [a]
5%	7.45 ± 0.02 [a]	0.23 ± 0.04 [a]	1.92 ± 0.03 [a]	14.68 ± 0.44 [a]	0.91 ± 0.08 [a]	39.6 ± 3.9 [a]	32.7 ± 0.4 [a]

EC: electrical conductivity; OC, organic carbon; CEC, cation exchangeable capacity. Mean values with the same letters were not significantly different ($p > 0.05$).

3.4. Tolerance Index and Bioavailability

Increasing the biochar rate increased the tolerance index at each Cd concentration level (Figure 4). The highest tolerance index was observed with B5 (Figure 4a). However, B1 and B3 also significantly improved the tolerance index, but 5% addition had the best effect at all levels of Cd pollution ($p < 0.05$). All biochar amendments decreased the bioavailability of Cd, which was most significant at biochar amendment rates of 3% and 5% and partly significant at 1% biochar application ($p < 0.05$) (Figure 4b). The lowest bioavailability of Cd was generally observed at a biochar application rate of 5% (Figure 4b). Generally, the bioavailability of Cd in all treatments increased with an increasing Cd level.

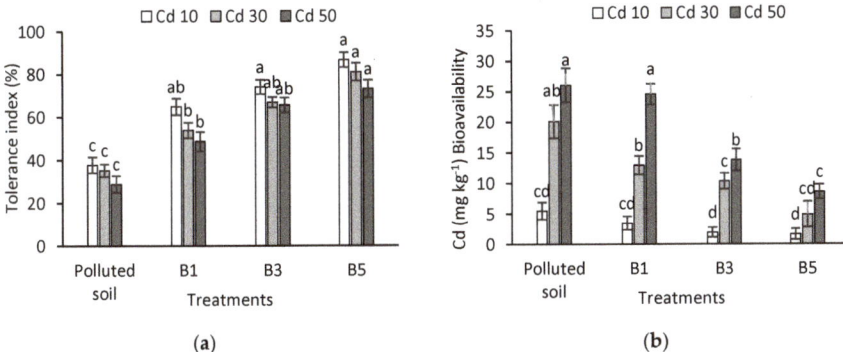

Figure 4. Effect of different levels of biochar on: (**a**) tolerance index and (**b**) bioavailability of Cd in oak seedlings at different levels of Cd contamination (means +/− one standard error). The term "Polluted soil" indicates pots with the Cd but without biochar. Different letters represent significant differences between treatments.

3.5. Cadmium in Plant Tissues

Higher soil Cd concentrations significantly caused greater Cd concentrations in all plant organs (Figure 5). However, using biochar amendments decreased the concentration of Cd in plant tissues significantly ($p < 0.05$) compared to the control at the same Cd pollution level which was the most pronounced in the leaves (Figure 5a) and the least pronounced (although still mostly significant) in the roots (Figure 5c). Again, B5 treatment caused the lowest concentration of Cd in plant tissues in all pollution levels of Cd.

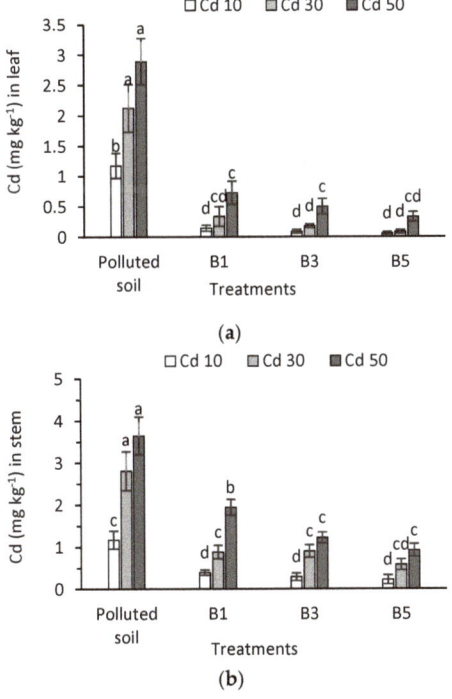

Figure 5. *Cont.*

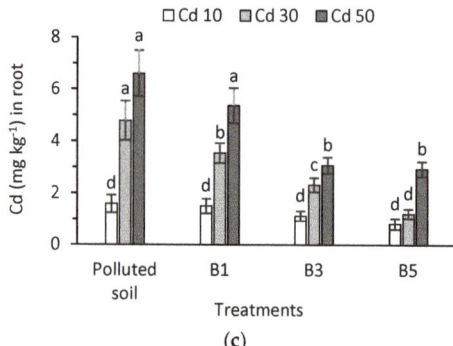

Figure 5. Effect of different levels of biochar on the concentrations of (a) Cd in leaves; (b) Cd in the stem; and (c) Cd in root tissue in oak seedlings at different levels of Cd contamination (means +/− one standard error). The term "Polluted soil" indicates pots with the Cd but without biochar. Different letters represent significant differences between treatments.

3.6. Cadmium Removal Efficiency

Increasing the amount of biochar amendment significantly decreased the Cd removal efficiency throughout all Cd pollution levels. The maximum removal efficiency for all Cd treatments was achieved with B5, while the minimum removal efficiency was observed with B1 (Figure 6).

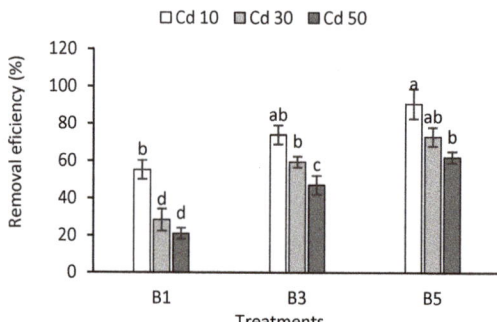

Figure 6. Effect of different levels of biochar on the removal efficiency in oak seedlings at different levels of Cd contamination (means +/− one standard error). Different letters represent significant differences between treatments.

With increasing biochar application level, Cd removal efficiency increased (Figure 7).

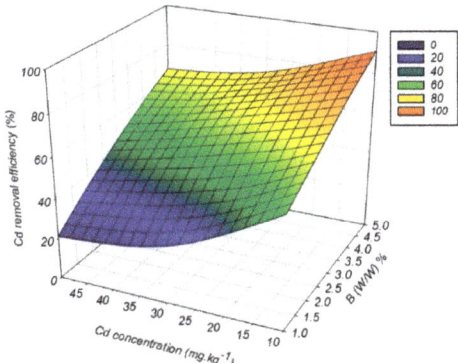

Figure 7. Cadmium removal efficiency as a function of Cd contamination level and biochar amendment level.

4. Discussion

As expected, increasing levels of Cd treatments significantly reduced plant diameter, height growth, and seedlings biomass (Figures 2 and 3). Kukier and Chaney [28] reported that Cd, by disrupting nutrient absorption and photosynthesis, reduces plant growth and biomass production. Similar results were reported by other researches [29,30].

In general, biochar had a positive effect on seedling growth and establishment, alleviating Cd-induced reductions in growth and development. Compared to biochar-free Cd-contaminated soils, biochar application increased seedling height and diameter, root, shoot, and leaf growth and hence the tolerance index mainly by reducing the amount of Cd that could be taken up by the seedlings [31]. Cation exchange capacity is one of the key factors in soil conditions that provides a high level of nutritional cations for plant use [6]. The natural CEC of the soil was 7.97 cmol$^{(+)}$ kg^{-1} before adding biochar (Table 2) and which was increased by adding B1, B3, and B5 to 12.31, 13.65, and 14.68 cmol$^{(+)}$ kg^{-1}, respectively (Table 4). This means an increase in the provision of cations in the soil for plant roots [32]. On the other hand, the high efficiency of biochar to adsorb metals and then increase plant growth [33] comes from its high porosity, a high number of functional groups, and high pH [34,35]. Also, the mineral components in biochar, including phosphates and carbonates, can precipitate with metals and reduce their bioavailability [36,37]. The results of the meta-analysis by Chen et al. [38] claimed that increases in the pH and CEC of soils are the main mechanisms of biochar in reducing metal bioavailability. The soil pH controls the solubility and bioavailability of the chemicals in the soil, both nutrients and pollutants, and therefore influences soil quality, crop productivity, and environment pollution [39]. In this study, the soil pH was slightly acidic (6.67). The bioavailability of micronutrients was higher than that of neutral-alkaline soils (soil pH > 7), enhancing crop productivity [40], but at the same time, the solubility of heavy metals and some of these nutrients may pass the critical level of environmental toxicity in these soils. Biochars, in general, have the potential and event priority to substitute lime for improving the properties of acidic soils and, therefore, enhanced plant growth [41]. Perhaps, the increase in pH (16%) and CEC (20%) compared to the control in the present investigation were related to the same reason.

Besides that, the use of biochar had positive effects on soil macronutrients (Table 4). Biochars, generally, are rich in plant nutrients (except for N when woody) and when added to soil, they usually increase soil fertility [30]. Masulili et al. [42] examined the effect of rice husk biochar on the soil properties of an acid sulfate soil and observed an increase in SOC, soil pH, CEC, N, P, K, and Ca and a decrease in exchangeable Al and soluble Fe. However, the amount of Mg and Na remained unaffected by biochar. Singh et al. [43] reported a significant increase in the water holding capacity, total C, N, and P, and soil moisture content.

According to the results, the addition of biochar significantly reduced the bioavailability of Cd in the soil (Figure 4), and the Cd concentrations in soil decreased with increasing biochar application levels (although not linearly). A reduced bioavailability of metals with biochar has been reported previously for Cd contamination in sandy soil to maize [44] and Cu toxicity in sandy soil to quinoa seedlings [45]. Several researchers [38,46,47] showed that biochar application reduced the concentration of metals and improved the condition of the plant and soil in terms of contamination because of its high surface area and CEC, as well as the presence of carboxyl, phenolic, hydroxyl, and other functional groups that contain superficial oxygen. On the other hand, the application of biochar often increases the pH value significantly, as shown by Lucchini et al. [48], and could hence decrease the mobility of the Cd.

The mechanism for this is that biochar increases the adsorption capacity of cations by increasing the negative charge on the soil surface [49–51]. Also, biochar is a porous substrate with a high surface area for the absorption of heavy metals and the formation of a complex with soluble organic carbon which immobilizes heavy metals [52]. According to Table 4, soil containing treatment B5 had the highest amount of organic carbon (1.92%). Similar results were reported by other studies [53,54].

Our findings show that the addition of biochar significantly reduced the concentration of Cd in the tissue of plants (Figure 5) as reported previously for maize by Liu et al. [3]. The low concentration of metals in the plant was directly related to the loss of metal bioavailability in the soil [38]. The pH of soil with the B5 treatment increased by 1.03 in the soil over the control. This increase may be related to pH = 8.14 in biochar. Li et al. [54] stated that the high pH of biochar is due to the dominance of carbonates and organic anions over ash. Generally, data on Cd accumulation in roots, stems, and leaves shows that Cd accumulates in the roots more than other plant organs (Figure 5), a strategy that was also employed by quinoa seedlings regarding Cu toxicity [45]. Due to the toxicity of Cd to cytosol in its free form, plant cells likely try to fix it at the root by ways such as attaching it to the cell wall [55], storing it in vacuoles [56], and chelating it with phytochelatin [55], thus reducing its toxicity. The result will be a low transfer of Cd to the upper plant organs.

As shown in Figure 6, increasing the level of biochar application decreased the cadmium concentration at all levels. Adding biochar to the soil will have a series of incremental consecutive consequences, including porosity, surface reactive sites, contact between Cd and biochar, the redox potential, and the removal efficiency of Cd [57] (Figure 7). The high CEC of biochar is directly related to its specific surface area and high porosity and therefore the CEC of soil can be increased (Table 4) [53]. The increase in negative charge due to the biochar oxidation is also a reason for the high biochar CEC [58].

5. Conclusions

In this study, biochar amendments of rice husk biochar clearly reduced the bioavailability of cadmium in increasingly contaminated soil, alleviating the negative impact of the Cd pollution on seedling physiology and growth. The plant tolerance index for the highest Cd rate of 50 mg kg^{-1} increased significantly by 40.9%, 56%, and 60.6% with rice husk biochar addition rates of 1%, 3%, and 5%, respectively. Therefore, biochar application in Cd-contaminated oak forests can be a viable solution for oak seedling establishment. It remains to be tested if contaminations with other cationic heavy metals will be likewise remediated by biochar use, but the encouraging results obtained in this study, combined with that of other studies [44,45], call for the real-world first field testing in contaminated mining soils in Iran.

Author Contributions: Conceptualization, E.A.; Methodology, S.M.H. and C.K.; Software, E.A. and M.G.; Validation, S.M.H. and C.K.; Formal Analysis, E.A. and M.G.; Investigation, E.A. and S.M.H; Resources, P.B.; Data Curation, E.A. and M.G.; Writing—Original Draft Preparation, E.A.; Writing—Review and Editing, E.A., S.M.H. and C.K.; Supervision, S.M.H.; Project Administration, E.A. All authors have read and agreed to the published version of the manuscript.

Funding: This research received no external funding.

Acknowledgments: The authors sincerely thank the Geisenheim University in Germany for providing the facilities and resources, e.g., extended literature analyses, for this research.

Conflicts of Interest: The authors declare no conflict of interest.

References

1. Seneviratne, M.; Weerasundara, L.; Ok, Y.S.; Rinklebe, J.; Vithanage, M. Phytotoxicity attenuation in Vigna radiata under heavy metal stress at the presence of biochar and N fixing bacteria. *J. Environ. Manag.* **2017**, *186*, 293–300. [CrossRef] [PubMed]
2. Al-Wabel, M.I.; Usman, A.R.; El-Naggar, A.H.; Aly, A.A.; Ibrahim, H.M.; Elmaghraby, S.; Al-Omran, A. Conocarpus biochar as a soil amendment for reducing heavy metal availability and uptake by maize plants. *Saudi J. Biol. Sci.* **2015**, *22*, 503–511. [CrossRef] [PubMed]
3. Liu, L.; Li, J.; Yue, F.; Yan, X.; Wang, F.; Bloszies, S.; Wang, Y. Effects of arbuscular mycorrhizal inoculation and biochar amendment on maize growth, cadmium uptake and soil cadmium speciation in Cd-contaminated soil. *Chemosphere* **2018**, *194*, 495–503. [CrossRef] [PubMed]
4. Soudek, P.; Valseca, I.R.; Petrová, Š.; Song, J.; Vaněk, T. Characteristics of different types of biochar and effects on the toxicity of heavy metals to germinating sorghum seeds. *J. Geochem. Explor.* **2017**, *182*, 157–165. [CrossRef]
5. Abbas, T.; Rizwan, M.; Ali, S.; Adrees, M.; Mahmood, A.; Zia-ur-Rehman, M.; Ibrahim, M.; Arshad, M.; Qayyum, M.F. Biochar application increased the growth and yield and reduced cadmium in drought stressed wheat grown in an aged contaminated soil. *Ecotoxicol. Environ.* **2018**, *148*, 825–833. [CrossRef]
6. Tekdal, D.; Cetiner, S. Investigation of the effects of salt (NaCl) stress and cadmium (cd) toxicity on growth and mineral acquisition of Vuralia turcica. *S. Afr. J. Bot.* **2018**, *118*, 274–279. [CrossRef]
7. Cui, L.; Pan, G.; Li, L.; Bian, R.; Liu, X.; Yan, J.; Quan, G.; Ding, C.; Chen, T.; Liu, Y.; et al. Continuous immobilization of cadmium and lead in biochar amended contaminated paddy soil: A five-year field experiment. *Ecol. Eng.* **2016**, *93*, 1–8. [CrossRef]
8. Lehmann, J.; Rillig, M.C.; Thies, J.; Masiello, C.A.; Hockaday, W.C.; Crowley, D. Biochar effects on soil biota: A review. *Soil. Biol. Biochem.* **2011**, *43*, 1812–1836. [CrossRef]
9. Mohamed, B.A.; Ellis, N.; Kim, C.S.; Bi, X. The role of tailored biochar in increasing plant growth, and reducing bioavailability, phytotoxicity, and uptake of heavy metals in contaminated soil. *Environ. Pollut.* **2017**, *230*, 329–338. [CrossRef]
10. Ippolito, J.A.; Laird, D.A.; Busscher, W.J. Environmental benefits of biochar. *J. Environ. Qual.* **2012**, *41*, 967–972. [CrossRef]
11. Ramzani, P.M.A.; Coyne, M.S.; Anjum, S.; Iqbal, M. In situ immobilization of Cd by organic amendments and their effect on antioxidant enzyme defense mechanism in mung bean (*Vigna radiata* L.) seedlings. *Plant. Physiol. Bioch.* **2017**, *118*, 561–570. [CrossRef] [PubMed]
12. Ghorbani, M.; Amirahmadi, E. Effect of rice husk Biochar (RHB) on some of chemical properties of an acidic soil and the absorption of some nutrients. *J. Appl. Sci. Environ. Manag.* **2018**, *22*, 313–317. [CrossRef]
13. Alobwede, E.; Leake, J.R.; Pandhal, J. Circular economy fertilization: Testing micro and macro algal species as soil improvers and nutrient sources for crop production in greenhouse and field conditions. *Geoderma* **2019**, *334*, 113–123. [CrossRef]
14. Fellet, G.; Marmiroli, M.; Marchiol, L. Elements uptake by metal accumulator species grown on mine tailings amended with three types of biochar. *Sci. Total Environ.* **2014**, *468*, 598–608. [CrossRef] [PubMed]
15. Sabeti, H. *Forests, Trees and Shrubs of Iran*; Ministry of Agriculture and Natural Resources: Iran, Tehran, 1976.
16. Tafazoli, M.; Hojjati, S.M.; Biparva, P.; Kooch, Y.; Lamersdorf, N. Reduction of soil heavy metal bioavailability by nanoparticles and cellulosic wastes improved the biomass of tree seedlings. *J. Plant. Nutr. Soil. Sci.* **2017**, *180*, 683–693. [CrossRef]
17. Zhang, Z.; Solaiman, Z.M.; Meney, K.; Murphy, D.V.; Rengel, Z. Biochars immobilize soil cadmium, but do not improve growth of emergent wetland species Juncus subsecundus in cadmium-contaminated soil. *J. Soil. Sediment.* **2013**, *13*, 140–151. [CrossRef]
18. Gamage, D.N.V.; Mapa, R.B.; Dharmakeerthi, R.S.; Biswas, A. Effect of rice-husk biochar on selected soil properties in tropical Alfisols. *Soil Res.* **2016**, *54*, 302. [CrossRef]

19. Ghorbani, M.; Asadi, H.; Abrishamkesh, S. Effects of rice husk biochar on selected soil properties and nitrate leaching in loamy sand and clay soil. *Int. Soil Water Conserv. Res.* **2019**, *7*, 258–265. [CrossRef]
20. Rajkovich, S.; Enders, A.; Hanley, K.; Hyland, C.; Zimmerman, A.R.; Lehmann, J. Corn growth and nitrogen nutrition after additions of biochars with varying properties to a temperate soil. *Biol. Fertil. Soils* **2012**, *48*, 271–284. [CrossRef]
21. Chan, Y.C.; Vowles, P.D.; McTainsh, G.H.; Simpson, R.W.; Cohen, D.D.; Bailey, G.M. Use of a modified Walkley-Black method to determine the organic and elemental carbon content of urban aerosols collected on glass fibre filters. *Chemosphere* **1995**, *31*, 4403–4411. [CrossRef]
22. Sumner, M.E.; Miller, W.P. Cation exchange capacity and exchange coefficients. *Methods Soil Anal. Part 3 Chem. Methods* **1996**, *5*, 1201–1229.
23. Helmke, P.; Sparks, D. Lithium, sodium, potassium, rubidium, and cesium. *Methods Soil Anal. Part 3 Chem. Methods* **1996**, 551–740. [CrossRef]
24. Olsen, S.R. *Estimation of Available Phosphorus in Soils by Extraction with Sodium Bicarbonate*; United States Department of Agriculture: Washington, DC, USA, 1954.
25. Quevauviller, P.; Rauret, G.; López-Sánchez, J.F.; Rubio, R.; Ure, A.; Muntau, H. Certification of trace metal extractable contents in a sediment reference material (CRM 601) following a three-step sequential extraction procedure. *Sci. Total. Environ.* **1997**, *205*, 223–234. [CrossRef]
26. Landberg, T.; Greger, M. Differences in oxidative stress in heavy metal resistant and sensitive clones of Salix viminalis. *J. Plant Physiol.* **2002**, *159*, 69–75. [CrossRef]
27. Poo, K.M.; Son, E.B.; Chang, J.S.; Ren, X.; Choi, Y.J.; Chae, K.J. Biochars derived from wasted marine macro-algae (Saccharina japonica and Sargassum fusiforme) and their potential for heavy metal removal in aqueous solution. *J. Environ. Manag.* **2018**, *206*, 364–372. [CrossRef]
28. Kukier, U.; Chaney, R.L. In situ remediation of nickel phytotoxicity for different plant species. *J. Plant Nutr.* **2004**, *27*, 465–495. [CrossRef]
29. Bolan, N.; Kunhikrishnan, A.; Thangarajan, R.; Kumpiene, J.; Park, J.; Makino, T.; Kirkham, M.B.; Scheckel, K. Remediation of heavy metal (loid) s contaminated soils–to mobilize or to immobilize? *J. Hazard. Mater.* **2014**, *266*, 141–166. [CrossRef]
30. Quintela-Sabaris, C.; Marchand, L.; Kidd, P.S.; Friesl-Hanl, W.; Puschenreiter, M.; Kumpiene, J.; Mueller, I.; Neu, S.; Janssen, J.; Vangronsveld, J.; et al. Assessing phytotoxicity of trace element-contaminated soils phytomanaged with gentle remediation options at ten European field trials. *Sci. Total. Environ.* **2017**, *599*, 1388–1398. [CrossRef]
31. Ramzani, P.M.A.; Iqbal, M.; Kausar, S.; Ali, S.; Rizwan, M.; Virk, Z.A. Effect of different amendments on rice (*Oryza sativa* L.) growth, yield, nutrient uptake and grain quality in Ni-contaminated soil. *Environ. Sci. Pollut. Res.* **2016**, *23*, 18585–18595. [CrossRef]
32. Nahar, K.; Rahman, M.; Hasanuzzaman, M.; Alam, M.M.; Rahman, A.; Suzuki, T.; Fujita, M. Physiological and biochemical mechanisms of spermine-induced cadmium stress tolerance in mung bean (*Vigna radiata* L.) seedlings. *Environ. Sci. Pollut. Res.* **2016**, *23*, 21206–21218. [CrossRef]
33. Xu, G.; Shao, H.B.; Sun, J.N. What is more important for enhancing nutrient bioavailability with biochar application into a sandy soil: Direct or indirect mechanism. *Ecol. Eng.* **2013**, *52*, 119–124. [CrossRef]
34. Zhang, X.; Wang, H.; He, L.; Lu, K.; Sarmah, A.; Li, J.; Huang, H. Using biochar for remediation of soils contaminated with heavy metals and organic pollutants. *Environ. Sci. Pollut. Res.* **2013**, *20*, 8472–8483. [CrossRef] [PubMed]
35. Wang, Y.; Wang, Z.L.; Zhang, Q.; Hu, N.; Li, Z.; Lou, Y.; Li, Y.; Xue, D.; Chen, Y.; Wu, C.; et al. Long-term effects of nitrogen fertilization on aggregation and localization of carbon, nitrogen and microbial activities in soil. *Sci. Total. Environ.* **2018**, *624*, 1131–1139. [CrossRef]
36. Cao, X.; Ma, L.; Gao, B.; Harris, W. Dairy-manure derived biochar effectively sorbs lead and atrazine. *Environ. Sci. Technol.* **2009**, *43*, 3285–3291. [CrossRef] [PubMed]
37. Kumar, M.; Rajput, T.B.S.; Kumar, R.; Patel, N. Water and nitrate dynamics in baby corn (*Zea mays* L.) under different fertigation frequencies and operating pressures in semiarid region of India. *Agric. Water Manag.* **2016**, *163*, 263–274. [CrossRef]
38. Chen, D.; Liu, X.; Bian, R.; Cheng, K.; Zhang, X.; Zheng, J.; Joseph, S.; Crowley, D.; Pan, G.; Li, L. Effects of biochar on availability and plant uptake of heavy metals–A meta-analysis. *J. Environ. Manag.* **2018**, *222*, 76–85. [CrossRef]

39. Weil, R.R.; Brady, N.C. *The Nature and Properties of Soils*, 15th ed.; Prentice Hall: Upper Saddle River, NJ, USA, 2016.
40. Lončarić, Z.; Karalić, K.; Popović, B.; Rastija, D.; Vukobratović, M. Total and plant available micronutrients in acidic and calcareous soils in Croatia. *Cereal. Res. Commun.* **2008**, *36*, 331–334.
41. Wu, S.; Zhang, Y.; Tan, Q.; Sun, X.; Wei, W.; Hu, C. Biochar is superior to lime in improving acidic soil properties and fruit quality of Satsuma mandarin. *Sci. Total. Environ.* **2020**, *714*, 136722. [CrossRef]
42. Masulili, A.; Utomo, W.H.; Syechfani, M. Rice husk biochar for rice based cropping system in acid soil 1. The characteristics of rice husk biochar and its influence on the properties of acid sulfate soils and rice growth in West Kalimantan, Indonesia. *J. Agric. Sci.* **2010**, *2*, 39. [CrossRef]
43. Singh, C.; Tiwari, S.; Gupta, V.K.; Singh, J.S. The effect of rice husk biochar on soil nutrient status, microbial biomass and paddy productivity of nutrient poor agriculture soils. *Catena* **2018**, *171*, 485–493. [CrossRef]
44. Namgay, T.; Singh, B.; Singh, B.P. Influence of biochar application to soil on the availability of As, Cd, Cu, Pb, and Zn to maize (*Zea mays* L.). *Soil Res.* **2010**, *48*, 638–647. [CrossRef]
45. Buss, W.; Kammann, C.; Koyro, H.W. Biochar reduces copper toxicity in Chenopodium quinoa Willd. in a sandy soil. *J. Environ. Qual.* **2012**, *41*, 1157–1165. [CrossRef] [PubMed]
46. Beesley, L.E.; Moreno-Jiménez, L.; Jose, J.L.; Gomez-Eyles, E.; Harris, B.; Robinson, B.; Sizmur, T. A review of biochars' potential role in the remediation, revegetation and restoration of contaminated soils. *Environ. Pollut.* **2011**, *159*, 3269–3282. [CrossRef] [PubMed]
47. Carter, S.; Shackley, S.; Sohi, S.; Suy, T.B.; Haefele, S. The Impact of Biochar Application on Soil Properties and Plant Growth of Pot Grown Lettuce (*Lactuca sativa*) and Cabbage (*Brassica chinensis*). *Agronomy* **2013**, *3*, 404–418. [CrossRef]
48. Lucchini, P.; Quilliam, R.S.; DeLuca, T.H.; Vamerali, T.; Jones, D.L. Does biochar application alter heavy metal dynamics in agricultural soil? *Agric. Ecosyst. Environ.* **2014**, *184*, 149–157. [CrossRef]
49. Lindsay, W.L. *Chemical Equilibria in Soils*; John Wiley & Sons Ltd.: Hoboken, NJ, USA, 1979.
50. McBride, M.; Sauve, S.; Hendershot, W. Solubility control of Cu, Zn, Cd and Pb in contaminated soils. *Eur. J. Soil Sci.* **1997**, *48*, 337–346. [CrossRef]
51. Ma, L.; Xu, R.; Jiang, J. Adsorption and desorption of Cu (II) and Pb (II) in paddy soils cultivated for various years in the subtropical China. *Int. J. Environ. Sci.* **2010**, *22*, 689–695. [CrossRef]
52. Shen, X.; Huang, D.Y.; Ren, X.F.; Zhu, H.H.; Wang, S.; Xu, C.; He, Y.B.; Luo, Z.C.; Zhu, Q.H. Phytoavailability of Cd and Pb in crop straw biochar-amended soil is related to the heavy metal content of both biochar and soil. *J. Environ. Manag.* **2016**, *168*, 245–251. [CrossRef]
53. Jien, S.H.; Wang, C.S. Effects of biochar on soil properties and erosion potential in a highly weathered soil. *Catena* **2013**, *110*, 225–233. [CrossRef]
54. Li, J.H.; Lv, G.H.; Bai, W.B.; Liu, Q.; Zhang, Y.C.; Song, J.Q. Modification and use of biochar from wheat straw (*Triticum aestivum* L.) for nitrate and phosphate removal from water. *Desalin. Water Treat* **2016**, *57*, 4681–4693.
55. Gong, J.; David, A.; Julian, I. Longdistance root-to-shoot transport of phytochelatins and cadmium in Arabidopsis. *Proc. Natl. Acad. Sci. USA* **2003**, *100*, 10118–10123. [CrossRef] [PubMed]
56. Lozano-Rodriguez, E.; Hernandez, L.E.; Bonay, P.; Carpena-Ruiz, R.O. Distribution of cadmium in shoot and root tissues of maize and pea plants: Physiological disturbances. *J. Exp. Bot.* **1997**, *48*, 123–128. [CrossRef]
57. Jiang, B.; Lin, Y.; Mbog, J.C. Biochar derived from swine manure digestate and applied on the removals of heavy metals and antibiotics. *Bioresour. Technol.* **2018**, *270*, 603–611. [CrossRef] [PubMed]
58. Mehmood, K.; Baquy, M.A.A.; Xu, R.K. Influence of nitrogen fertilizer forms and crop straw biochars on soil exchange properties and maize growth on an acidic Ultisol. *Arch. Agron. Soil. Sci.* **2018**, *64*, 834–849. [CrossRef]

© 2020 by the authors. Licensee MDPI, Basel, Switzerland. This article is an open access article distributed under the terms and conditions of the Creative Commons Attribution (CC BY) license (http://creativecommons.org/licenses/by/4.0/).

Article

Assessment and Mitigation of Heavy Metals Uptake by Edible Vegetables Grown in a Turin Contaminated Soil Used as Vegetable Garden

Elisa Gaggero [1,*], Paola Calza [1], Debora Fabbri [1], Anna Fusconi [2], Marco Mucciarelli [2], Giorgio Bordiglia [1], Ornella Abollino [3] and Mery Malandrino [1]

[1] Department of Chemistry, University of Torino, Via Pietro Giuria 5, 10125 Torino, Italy; paola.calza@unito.it (P.C.); debora.fabbri@unito.it (D.F.); giorgio.bordiglia@gmail.com (G.B.); mery.malandrino@unito.it (M.M.)
[2] Department of Life Sciences and Systems Biology, University of Torino, Via Accademia Albertina 13, 10123 Torino, Italy; anna.fusconi@unito.it (A.F.); marco.mucciarelli@unito.it (M.M.)
[3] Department of Drug Science and Technology, University of Torino, Via Pietro Giuria 9, 10125 Turin, Italy; ornella.abollino@unito.it
* Correspondence: elisa.gaggero@unito.it

Received: 1 June 2020; Accepted: 25 June 2020; Published: 29 June 2020

Abstract: In this study we evaluated the concentration of 22 elements, namely Al, As, Ba, Ca, Cd, Ce, Co, Cr, Cu, Fe, K, La, Mg, Mn, Na, Ni, P, Pb, Sr, Ti, V, Zn, and their uptake by edible plants in soils collected in a green urban area. The results highlighted a high yield of those heavy metals typical for anthropic pollution, such as Pb, Zn, Cu, Ba and Co, likely due to the intensive vehicular traffic. The uptake of metals by edible plants was analyzed on two broadleaf plants, *Lactuca sativa* and *Brassica oleracea*, grown in this soil and in an uncontaminated Turin soil in a growth chamber with and without the addition of a soil improver, provided by a local Organic Waste Treatment Plant. The subsequent analysis of their aerial part and roots highlighted the absorption of the main potentially toxic elements (PTEs) by the vegetables grown in the contaminated soil, whereas their concentration was lower if cultivated in the comparison soil, which was free of pollutants. The use of a soil amendment did not decrease the uptake of PTEs by *Lactuca sativa* and *Brassica oleracea*, but it caused a strong reduction in their translocation from the roots to the edible part, which consisted of the aerial part.

Keywords: urban gardens; pollution; heavy metals; contaminated soil; amendment; edible plants

1. Introduction

In recent years, the urban and peri-urban horticulture underwent a widespread diffusion in European cities prompted by the need for more sustainable and inclusive growth. Urban gardens are green spaces, owned by the municipal administrations, rented to associations, groups or individuals for the cultivation of flowers, fruit and vegetables [1]. Often, they are realized in peripheral areas of the city, constituting transition areas between the city and the countryside at risk of building speculation.

Urban gardens can be considered a concrete response to needs related to the concept of common good [2]. They represent an ancient, but highly innovative system to improve access to food for citizens, mainly for the poorest population, and to reduce socio-economic and environmental issues. The interest in urban horticulture is due to the need for citizens to regain green areas in contrast with increasing urbanization, the greater attention to fresh products with a short food supply chain and to an enhanced sensitivity toward sustainability. In fact, this practice favors an environmentally friendly urban regeneration that implies a reduction in the energy impact of the cities, an improvement in the urban microclimate, an overbuilding limitation, a sustainable cultivation by fighting food waste, a diffusion of virtuous circuits of fair trade, and ultimately a local production system through micro-economy paths [3,4]. It also contributes to increased urban life quality thanks to the aesthetic satisfaction given by a greener urban environment [5]. Moreover, the social impact of this activity is extremely positive since this community experience brings together different generations and different social contexts and encourages social aggregation, constituting a place for meeting and sharing. For instance, it has an important role in the rehabilitation of people with alcohol and drug addictions, for supporting and helping the elderly or physically and mentally disabled people [6].

Turin can be considered the Italian capital of urban gardens, due to the attention it has always paid to their planning and implementation. Currently the city counts 15 regulated urban gardens (total area of approximately 64,232 m^2), 22 associative gardens, 7 spontaneous urban gardens and 3 vegetable gardens under construction [7].

Nevertheless urban gardens, and hence the plants growing therein, are exposed to continuous pollution sources such as atmospheric deposition, polluted irrigation water, vehicular traffic, use of pesticides and, mainly in post-industrial cities like Turin, previous industrial settlements [8]. Therefore, pollutants can be absorbed by crops, leading to a potential risk for human health. In particular, exposure to heavy metals is the most important risk factor in urban areas, especially as regards lead [9], whose adverse effects on health have been extensively documented [10].

For these reasons, it is imperative to determine the level of heavy metals pollution in the soil envisioned for the cultivation of edible species and to evaluate the yield of inorganic contaminants absorbed by plants. It is also fundamental to investigate a possible correlation between the content of contaminants in the edible plant species and that of the area where the plant grew. In plants, both the bioaccumulation factor and the translocation factor from root to shoot greatly vary with the plant species, the type of heavy metal and their interactions Thus, the risk of human exposure to contaminants can be significantly reduced by selecting suitable plants.

Generally, broad-leaved plants, such as species belonging to the genus *Brassica*, are more liable to store heavy metals in their edible parts [11–13], and the genus *Lactuca* exhibits a great absorption capacity for different metals [13]. Toxic elements present in the soil are usually absorbed by the plant at a radical level through binding carrier proteins normally involved in the transport of nutrients [12,14], and by acidifying the rhizosphere through proton pumps, with the consequent release of low-molecular-weight compounds which are able to act as metals chelators, thus favoring their uptake [14].

Our study aims to assess the level of heavy metal contamination in an urban soil in Turin, suitability for its position as an urban garden, its effects on the growth and uptake capacity of two selected broad-leaved plants, *Lactuca sativa* and *Brassica oleracea*, usually found in vegetable gardens.

2. Materials and Methods

2.1. Soil Sampling and Characterization

A sandy-loam soil was collected in a garden situated in a central area of Turin, in San Salvario neighborhood (CA-Via F. Campana, GPS coordinates: N 45°3'14", E 7°40'57") which is supposed to have a high level of heavy metals content. For comparison, another Turin sandy-loam soil, known as uncontaminated by heavy metals, was collected in a hilly peripheral green area (NOB-Parco del Nobile, GPS coordinates: N 45°3'1", E 7°42'52").

As regards the CA site, a more thorough sampling was carried out in February 2018 considering several points both on the surface and on the deep layer. The site was divided into four cells of 4 m^2 area following a square grid layout, named A, B, C, D. The soil collected at the surface (10 cm) of vertexes and in the center of each square cell was mixed to obtain composite samples, namely SurA, SurB, SurC and SurD. The samples DeepA and DeepB were collected at a depth of 40 cm in the center of cells A and B, respectively. In the NOB site, five cores were randomly collected within a 40 m^2 area. During the procedure, plastic tools were used in order to avoid contaminations and the samples were stowed into polyethylene bags. Thereafter, they were air-dried, sieved on a 2 mm sieve, ground in a centrifugal ball mill and stored in carbonate jars until they were analyzed.

2.2. Pot Experiments

Pots with a capacity of 600 mL were filled with the soil taken from the surface in the two sites after placing at the bottom a fine net of 10 cm^2 and a layer of coarse sand (sterilized in a stove at 150 °C) to facilitate drainage.

Lactuca sativa L. and *Brassica oleracea* L. var. *capitata* were grown in pots, under controlled conditions and they were cultivated with and without the addition of a soil amendment supplied by Acea Pinerolese (AM) to evaluate its effect on the plants. This concentrated soil amendment comprises composted green waste and anaerobically digested organic materials from several sources. The composting process consists of two phases, namely the bio-oxidation (active composting time) of the most easily degradable organic components and the maturation (curing phase) in which the product is stabilized and enriched with humic substances [15]. The final product has the following characteristics: pH = 7–8.8, humidity = 50%, C/N ratio = 25, organic nitrogen = 80% NTK, total organic carbon = 20% p/p, electrical conductivity = 1.8 mS/cm. *Lactuca sativa* and *Brassica oleracea* were sown and grown in CA and in NOB soils both unaltered and mixed with the soil improver (70% soil and 30% soil improver (v/v)). Five pots were prepared as replicates for each plant for a total of 20 pots, considering the two soil sampling sites and the treatment with and without the amendment. They were kept in a growth chamber under a 16/8 h photo- and thermo-period of 24/20 °C day/night. The light source consisted of four Green Power led production modules (Philips) per shelf with a 2:1 red (626 nm peak) and blue (470 nm peak) emission. The plants were watered three times a week with 50–100 mL of deionized water for each watering.

The plants were harvested 3 months from sowing, and morphometric measurements were made on the diameter at the base of the stem and the number and area of the leaves. For the latter, the most representative leaves were scanned and the area was calculated through ImageJ, a computer program for digital image processing [16].

The root system was washed and separated from the rest of the plant. Fresh and dry weights were calculated before and after a drying process at a temperature of 60 °C in the oven until a constant weight was reached.

2.3. Treatment and Analysis of Vegetable and Soil Samples

Soil organic carbon (SOC), organic matter (SOM) and pH were measured in the soil of both sites according to the Italian official soil chemical analysis methods described by the Ministerial Decree 13/9/1999 [17]: pH was determined by potentiometric analysis on soil-CaCl$_2$ suspension, SOC and SOM were measured with the Walkley–Black method [18].

The aerial part and roots were freeze-dried and ground to obtain a sample as homogeneous as possible. Before proceeding with the analysis, the samples of soil and vegetables were pre-treated mineralizing them with a Milestone MLS-1200 Mega (Milestone, Sorisole, Italy) microwave.

The digestion procedure for soil samples was performed by adding 5 mL of aqua regia to 0.05 g of soil inside tetrafluoromethoxyl (TFM) vessels, whereas for vegetables samples 0.5 g of epigeal or hypogeal parts were weighed and added with 6 mL of nitric acid and 2 mL of hydrogen peroxide. This mixture is strongly oxidizing and develops a lot of foam; for this reason, the vessels were let to rest under the hood for about thirty minutes before starting the digestion. Three replicates were made for each soil and vegetable sample. Soil and plant extracts, obtained from the mineralization step, were filtered on a Whatman 40 paper filter and then diluted with Highly Purified Water (HPW) to a final volume of 50 mL.

Twenty elements (Al, As, Ba, Ca, Cd, Ce, Co, Cr, Cu, Fe, K, La, Mg, Mn, Na, Ni, P, Pb, Sr, Ti, V, Zn) in plant samples and twenty-two (with the addition of La and Ce) in soil samples were determined by inductively coupled plasma–optical emission spectrometer (ICP-OES) or, when below the detection limit (LOD) of this instrument, the chemical analyses were conducted by a high-resolution inductively coupled plasma-mass spectrometer (HR-ICP-MS) or a graphite furnace atomic absorption spectrometer (GF-AAS), according to the different matrixes of soil and vegetables samples. In particular, As and Cd were determined by GF-AAS in all soil samples since they presented concentrations below the LOD of ICP-OES. As regards the plant samples, Cu, Ti and Al were below the LOD of ICP-OES and were determined by HR-ICP-MS in the 25% of samples, Cr, Ni and V in the 50% and Pb and Co in the 75% of samples. The other elements were quantified using the ICP-OES. Three instrumental replicates were performed for the determination of each element. The quality of the analytical procedure was verified using two NIST standard reference materials: tomato leaves 1573a for plant samples and San Joaquin soil 2709 for soil samples.

Models and technical specification of instruments are reported in Table 1.

Table 1. Models and technical specification of instrumental techniques used for the analyses.

Technique	Model	Features
ICP-OES	Perkin Elmer Optima 7000 DV	Mira Mist nebulizer, cyclonic spray chamber, dual echelle monochromator, dual CCD detector
HR-ICP-MS	Thermo Finnigan Element 2	Conical nebulizer, cyclonic spray chamber, magnetic and electric sector, SEM detector
GF-AAS	Perkin Elmer Aanalyst 600	Transversely Heated Graphite Atomizer (THGA) furnace assembly, longitudinal Zeeman-effect background correction, enhanced STPF technology, True Temperature Control (TTC), solid-state detector

Calibration curves were performed using the external standard calibration method for the quantification of all the elements. Standard solutions were prepared by diluting single element concentrated (1000 mg L^{-1}) stock solutions (Sigma-Aldrich TraceCERT), with aliquots of sample blanks prepared with the same acid mixtures of samples (matrix matching method) [19].

The bioconcentration factor for each vegetable was calculated to assess the yield of metals transferred from the soil to the plant as follows [20,21]:

BF = metal concentration in aerial part and roots (mg kg^{-1} dry weight)/metal concentration in soil (mg kg^{-1} dry weight).

The translocation factor allows evaluating to what extent elements were transferred from the root system to the aerial part of the plant and was calculated for each vegetable through the relationship [20,21]:

TF = metal concentration in aerial part (mg kg^{-1} dry weight)/metal concentration in roots (mg kg^{-1} dry weight).

2.4. Data Analysis

Data about plants' weights and morphometric parameters and the experimental results obtained by chemical analyses were processed by statistical treatment using the XlStat 2017 software package, an add-on of Microsoft Excel. Analysis of variance (ANOVA) and Tukey test with a level of confidence of 95% were performed [22,23].

3. Results and Discussion

3.1. Soil Characterization

The values of SOC, SOM and pH for the investigated soils are shown in Table 2.

Analyses of CA soil evidence higher contents of SOC and SOM in surface layer than on deeper layers. These results are not surprising, as they are generally higher in the first decimeters of the soil [24]. SOC and SOM content is grater for NOB soil compared to CA soil. SOC accomplishes an essential positive function on many soil properties; it promotes the aggregation and stability of soil particles and it binds effectively with numerous substances, improving soil fertility, microbial activity, and nutrient availability such as nitrogen and phosphorus [25]. SOM is largely constituted by high-molecular-weight organic materials such as polysaccharides, proteins, sugars, amino acids and humic substances. Humic and fulvic acids can be present in a dissociated form, and thus are negatively charged (the main sources of these charges are carboxylic and phenolic groups in which hydrogen can be replaced by metal ions). The source of negative charges in soil colloids is strongly pH-dependent, and so the sorption of heavy metals in soils with relatively high organic content is strongly affected by pH as well as the ion properties of metals, like charge and ionic radius [26]. Moreover, the pH influences the availability of the nutrients for plants that reaches the maximum in the range 6–7 for P, 6.5–8 for macronutrients (K, Ca, Mg), and 5–7 for micronutrients such Cu, Fe, Mn, Ni, and Zn. Therefore, neutral conditions appear to be optimal for crop growth and soil microorganisms' activity [27]. In our case, pH is close to neutrality for both terrains. Considering the sandy-loam texture of soils, the medium-high percentages of organic matter and organic carbon [25], which foster metals immobilization, and the neutral pH, marked metal availability and massive uptake by cultivated plants were not expected [28,29].

Table 2. General characteristics of the Turin central area (CA) and peripheral green area (NOB) soils.

	CA (Surface Layer)	CA (Deep Layer)	NOB
Organic carbon (% w/w)	1.66 ± 0.01	0.47 ± 0.02	2.05 ± 0.05
Organic matter (% w/w)	2.86 ± 0.01	0.80 ± 0.03	3.54 ± 0.08
pH	7.1	7.0	7.3

The concentration of the 22 elements detected in the superficial and deep layer samples of CA site, as well as their mean values and limits set by Italian legislation [30] are reported in Table 3. Overall, high concentration of Cr, Zn, Ni, Pb and Co were found in the CA site that exceeded public green area limits but not industrial area limits. NOB soil presents concentration values below the legal limits for all elements, except for Co, Cr and Ni.

It can be noticed that the content of some elements, namely Ba, Cd, Cu, P, Pb and Zn, was higher on the surface soil layer. The higher yield of P on the surface is not surprising as it is derived from decomposed organic material (leaves, insects) or from the addition of organic soil amendments or fertilizers [31]. The other results could be attributable to anthropogenic contamination, as these metals are commonly linked to urban pollution. For instance, Ba may derive from its widespread use in manufactured materials such as tiles, automobile clutch and brake linings, rubber, brick, paint, glass, and other materials [32]. Cd input in the environmental is due to different sources, such as vehicular traffic, combustion of fossil fuels and various industrial activities including the production of steel, pigments and batteries.

As concerns Pb, this metal is used in manifold activities, e.g., the production of automotive batteries, the manufacture of metal alloys, the working of the crystal, the welding and the production of bullets [33,34] and, in the past, in the production of paints and enamels, pipes, pesticides and as a gasoline additive for use as an anti-knock agent [35].

The anthropogenic sources of Cu should come from its use in agriculture (pesticides, agronomic use of zootechnical sewage, sewage sludge and amendment) and in industry, whereas the main sources of Zn are the industries (electroplating, foundries, battery production, etc.) and road traffic [36–38].

However, elements such as As, Cd and Cu, although more concentrated in the topsoil and therefore deserving of monitoring, the presented values below the Italian legal limits. Pb and Zn, instead, showed concentrations well above the limit for public green areas and their presence is probably connected to vehicular traffic, which represents the primary pollution source for the site of interest.

Table 3. Concentrations (mean and standard deviation) of the elements determined in the soil samples collected in the CA site and comparison with Italian legislation limits for green public areas (limit A) and industrial areas (limit B). All the concentrations are expressed as mg kg^{-1} dry weight. <LOD indicates that the concentration of the element is below the limit of detection of the analytical technique. Element concentrations that exceed limits for green public areas are reported in bold.

	Sup A			Sup B			Sup C			Sup D			Prof A			Prof B			Surficial Samples Mean Value			Deep Samples Mean Value			Limit A	Limit B
Al	44,372	±	1767	45,331	±	1100	41,785	±	3305	42,421	±	2212	52,790	±	1958	51,645	±	1625	43,477	±	1655	52,218	±	810		
As	10.9	±	0.3	14	±	1	13	±	2	16	±	4	9.6	±	0.2	10.0	±	0.4	13	±	2	9.8	±	0.3	20	50
Ba	396	±	56	472	±	31	467	±	92	461	±	93	229	±	4	232	±	6	449	±	35	231	±	2		
Ca	18,167	±	903	20,269	±	601	21,274	±	3424	20,803	±	1995	33,281	±	2152	30,341	±	2129	20,128	±	1370	31811	±	2079		
Cd	0.25	±	0.03	0.28	±	0.04	0.18	±	0.02	0.17	±	0.03	<LOD			<LOD			0.22	±	0.05	<LOD			2	15
Ce	40	±	3	45	±	5	42	±	7	41	±	6	52	±	5	49	±	6	42	±	2	50	±	3		
Co	**25**	±	**3**	**25**	±	**5**	**26**	±	**6**	19	±	6	**22**	±	**5**	**23**	±	**1**	**24**	±	**3**	**22.3**	±	**0.3**	20	250
Cr	**191**	±	**2**	**252**	±	**66**	**211**	±	**52**	**254**	±	**76**	**233**	±	**23**	**210**	±	**21**	**227**	±	**31**	**222**	±	**17**	150	800
Cu	89	±	17	89	±	3	79	±	3	91	±	22	58	±	1	59	±	2	87	±	6	59	±	1	120	600
Fe	37,379	±	9710	36,442	±	961	34,696	±	417	37,475	±	1649	35,779	±	535	36,379	±	220	36,498	±	1289	36,079	±	424		
K	11,804	±	500	12,316	±	363	11,343	±	699	11,858	±	233	12,865	±	409	12,932	±	375	11,830	±	398	12,899	±	47		
La	15	±	2	17	±	2	16	±	3	16	±	2	20	±	2	19	±	2	16	±	1	19.4	±	0.7		
Mg	12,313	±	639	13,357	±	738	12,535	±	1915	13,688	±	439	15,404	±	1334	14,811	±	1872	12,973	±	655	15,108	±	419		
Mn	866	±	61	941	±	10	936	±	7	928	±	34	980	±	12	1036	±	33	918	±	35	1008	±	40		
Na	7034	±	342	7762	±	110	8254	±	819	7566	±	537	8011	±	97	8518	±	19	7654	±	505	8265	±	359		
Ni	**142**	±	**12**	**167**	±	**26**	**157**	±	**26**	**159**	±	**7**	**139**	±	**4**	**148**	±	**10**	**156**	±	**10**	**143**	±	**6**	120	500
P	999	±	141	1407	±	258	1196	±	91	1013	±	19	383	±	11	403	±	15	1154	±	191	393	±	15		
Pb	**267**	±	**68**	**302**	±	**17**	**249**	±	**21**	**322**	±	**137**	57.0	±	1.4	56.3	±	1.4	**285**	±	**33**	56.7	±	0.5	100	1000
Sr	92	±	3	104	±	12	94	±	14	93	±	4	115	±	6	117	±	7	96	±	5	116	±	1		
Tl	2730	±	419	2797	±	149	2731	±	67	2633	±	200	2974	±	176	3235	±	152	2723	±	68	3105	±	185		
V	77	±	1	84	±	2	82	±	4	79	±	4	79.5	±	0.6	85	±	2	81	±	3	82	±	4	90	250
Zn	**220**	±	**12**	**257**	±	**11**	**225**	±	**8**	**218**	±	**8**	95	±	3	98	±	2	**230**	±	**18**	97	±	2	150	1500

On the contrary, Co, Cr and Ni concentration is almost the same in surface and deep samples, because their presence in the soils of Turin area is mainly attributable to the lithological substrate of ultramafic rock of which they are the main constituents, as documented by other studies on the soil metal content in the Turin area [39,40].

3.2. Pot Experimets

Morphometric measurements and analysis of metal content were carried out on the *L. sativa* and *B. oleracea* plants grown in the growth chamber with different soil samples in the presence or absence of a soil amendment.

Morphological measurements are collected in Table 4 and show that both the type of soil and the addition of a soil amendment influenced the biomass of *B. oleracea* and *L. sativa* as fresh weight of aerial part and roots, as confirmed by ANOVA and Tukey tests (Table S1, Supplementary Materials).

Table 4. Weights observed for *Lactuca sativa* and *Brassica oleracea* plants in the four growing conditions after three months of growing.

Vegetable Species	Soil	Amendment (AM)	Aerial Part Fresh Weight [g]	Root Fresh Weight [g]	Root Fresh Weight/Aerial Part Fresh Weight [g]	Aerial Part Dry Weight/Aerial Part Fresh Weight [g]
B. oleracea	NOB	No	6.6 ± 0.7	1.5 ± 0.4	0.22 ± 0.03	0.36 ± 0.06
		Yes	60.9 ± 7	5 ± 1	0.09 ± 0.01	0.25 ± 0.04
	CA	No	5.0 ± 0.5	1.4 ± 0.4	0.3 ± 0.1	0.23 ± 0.03
		Yes	28 ± 2	4 ± 1	0.14 ± 0.04	0.19 ± 0.03
L. sativa	NOB	No	15 ± 2	4 ± 1	0.24 ± 0.07	0.11 ± 0.01
		Yes	113 ± 11	10 ± 3	0.09 ± 0.02	0.06 ± 0.01
	CA	No	5.32 ± 0.06	1.5 ± 0.5	0.28 ± 0.07	0.14 ± 0.03
		Yes	37 ± 5	8 ± 3	0.20 ± 0.05	0.07 ± 0.01

In the absence of the amendment, the difference between the fresh weight of the aerial part of the plants grown in the NOB soil and in the CA soil is remarkable for *L. sativa*, whereas the values concerning *B. oleracea* do not differ significantly. Adding the amendment, a relevant increase in weight for both plants in CA soil and even more in NOB soil can be noted.

The root system shows the same trend. The ratio between the fresh weight of roots and the aerial part was higher in the plants grown in the absence of an amendment in both types of soil, since in this condition plants develop a more extensive root system due to the reduced availability of macro- and micro-nutrients compared with the soil with AM added to it. Considering the ratio between the dry and fresh weights of the aerial part of *B. oleracea*, a significant difference between the values of the plants grown in unpolluted soil without soil amendment and all other conditions can be observed, whereas higher values were found in both soils for *L.sativa* grown in the absence of AM. These data indicate a higher water content for plants grown in soils added with the amendment, which acts positively by reducing stomatal resistance.

The data reported in Table 5 and Figure 1 and ANOVA and Tukey tests (Table S1, Supplementary Materials) highlight that the soil used for cultivation particularly influenced the leaf area of both plants, showing the highest values for NOB soil. The addition of AM significantly affects the plant growth, as higher values of stem diameter and of number and size of the leaves are measured in all the examined plants for both sites. In particular, the stem diameter of *B. oleracea* and *L. sativa* grown with AM are about three and two times those of the plants grown without and the number of leaves is about double in the presence of AM for both plants.

To assess the adsorption in investigated edible plants of potentially harmful metals, the content of 22 elements was determined in the aerial parts and roots (data reported in Tables S2–S5, Supplementary Materials). Figure 2 shows the concentrations of potentially toxic elements (As, Ba, Cd, Co, Cr, Cu, Ni, Pb and Zn) analyzed in the roots (a, b) and in the aerial parts (c, d) of *L. sativa* grown in the soil of the two sites with or without the addition of AM, whereas the results of ANOVA and Tukey tests are reported in Tables S6 and S7 of Supplementary Materials. In general, the metal content in any part of the *L. sativa* is lower in those grown in the NOB soil, reflecting the difference in concentration detected in the two sites, except for the Cd, which shows similar concentrations.

Table 5. Morphometric parameters relating to *Lactuca sativa* and *Brassica oleracea* plants grown for 3 months under the investigated experimental conditions.

Vegetable Species	Soil	Amendment (AM)	Stem Diameter [mm]	Leaves Number	Leaf Area [cm^2]
B. oleracea	NOB	No	2.2 ± 0.1	8 ± 1	24 ± 4
		Yes	6.3 ± 0.6	12 ± 1	112 ± 9
	CA	No	1.8 ± 0.2	6 ± 1	8 ± 1
		Yes	5 ± 1	12 ± 2	50 ± 10
L. sativa	NOB	No	7.21 ± 0.9	16 ± 2	24 ± 1
		Yes	14 ± 3	33 ± 2	86 ± 3
	CA	No	6.8 ± 0.5	14 ± 1	14.1 ± 0.3
		Yes	10.9 ± 0.1	31 ± 2	57 ± 3

Figure 1. Plants of *Lactuca sativa* (top) and *Brassica oleracea* (bottom) grown under different conditions: (a) = NOB soil with AM, (b) = NOB soil without AM, (c) = CA soil with AM, (d) CA soil without AM.

The effectiveness of the addition of soil amendment depends both on the type of plant and the soil, and on the element under examination, which can have a different speciation and reactivity.

Considering both the root and aerial part of *L. sativa* grown in CA soil, it can be noticed that the total concentration of Ba, Co, Cr, Ni and Pb is greater for the vegetable grown in the presence of AM, but there is a minor translocation from the roots to the aerial part, with a consequently lower concentration in the edible part when the amendment is used. The plant is probably positively affected by the presence of the amendment and develops a more efficient root system that allows it to absorb a greater quantity of elements; however, these are not transferred to the edible aerial part. In particular, Ba and Ni adsorption seems to be significantly affected by the addition of AM since the difference between their content in the *L. sativa* samples grown in soils treated or not is very marked. In fact, the addition of the amendment causes a decrease in the concentration of Ni and Ba in the aerial parts of the plants grown in CA soil. Zinc, similar to the elements mentioned so far, presents a higher total concentration in *L. sativa* grown in the presence of AM, but in this case the amendment does not cause a strong reduction in its transfer to the edible part. Therefore, Zn, an essential microelement for plant life and very abundant in the soils considered, is well absorbed by the plant.

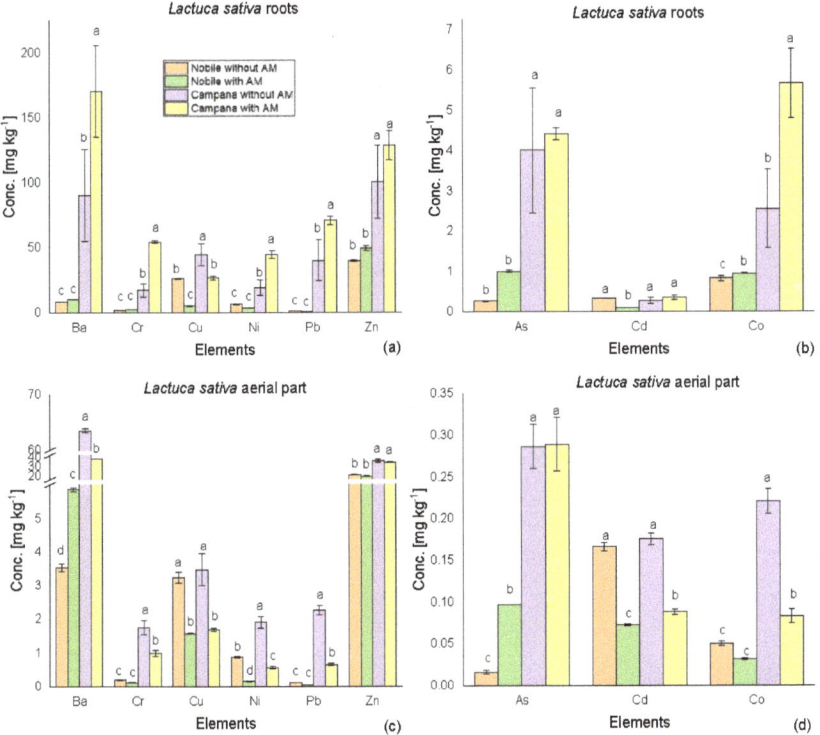

Figure 2. Comparison of potentially toxic elements' uptake in the roots (**a,b**) and in aerial part (**c,d**) of *Lactuca sativa* grown in CA and NOB soils. Elements concentrations are expressed as mg kg^{-1} dry weight. Letters above columns refer to the summary of multiple comparisons for pairs obtained with Tukey test. Letter **b** identifies plant samples whose mean concentrations in the respective elements are significantly lower than the ones obtained for plant samples identified by letter **a**; the same concept applies to letter **c** with respect to letter **b** and letter **d** with respect to letter **c**.

Copper is also an essential element for vegetable species, but its behavior is different since its uptake in the *L. sativa* is strongly influenced by the presence of amendment in both soils. Copper amount is always higher in *L. sativa* cultivated without amendment and this could be due to the different speciation of this element, often added as fungicide to the plants also cultivated in urban gardens, and at the lowest concentration necessary for plant growth.

In the case of As and Cd, the total concentration in the vegetable is comparable. The transport of Cd in the aerial parts of the plant is greater for the samples grown without the use of amendment, contrary to what was observed for As, where the addition of amendment does not cause a decreased concentration in the aerial part.

Regardless of the improver's use, the concentration values of Pb and Cd in *L. sativa* (Table S2) are well below the maximum concentration allowed for edible parts of broad-leaved vegetables by Commission Regulation (EC) No. 1881/2006 of the Official Journal of the European Union [41], which defines as a threshold 0.30 mg kg^{-1} wet weight for Pb and 0.20 mg kg^{-1} wet weight for Cd.

In Figure 3, the concentrations of potentially toxic elements analyzed in roots (a, b) and in the aerial part (c, d) of *B. oleracea* grown in the soil of the two sites with or without the addition of the soil amendment are reported, whereas the results of ANOVA and Tukey tests are reported in Tables S8 and S9 of Supplementary Materials.

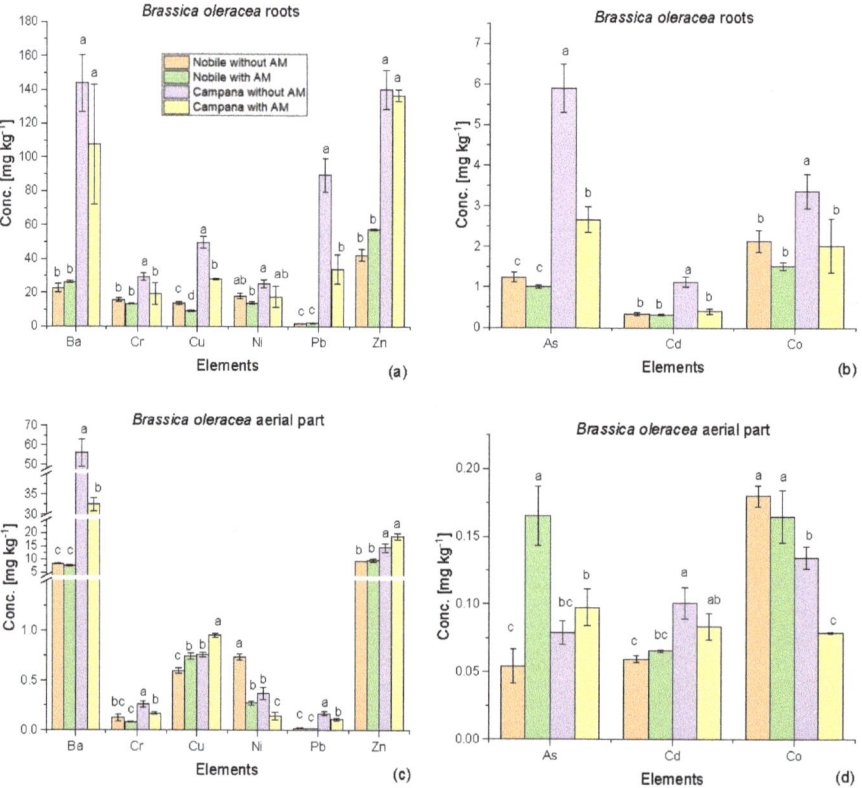

Figure 3. Comparison of potentially toxic elements' uptake in the roots (**a**,**b**) and in the aerial part (**c**,**d**) of *Brassica oleracea* grown in CA and NOB soils. Element concentrations are expressed as mg kg^{-1} dry weight. The letters above columns refer to the summary of multiple comparisons for pairs obtained with Tukey test. Letter **b** identifies plant samples whose mean concentrations in the respective elements are significantly lower than the ones obtained for plant samples identified by letter **a**; the same concept applies to letter **c** with respect to letter **b** and letter **d** with respect to letter **c**.

As in the case of *L. sativa*, the total metal content is higher for *B. oleracea* grown in the CA soil than for the vegetable grown in NOB soil, reflecting the different metal content of the two soils. The concentration gap is particularly significant in the case of the roots, while it decreases if the aerial part is considered. In particular, Ni, As and Co show unusually higher concentrations in the aerial part of *B. oleracea* grown in NOB soil.

Chromium, copper, lead, barium, cobalt, arsenic, and cadmium are more concentrated in the roots of *B. oleracea* grown in the polluted soil of CA site and in the absence of the AM for both soils. As regards the aerial part, the use of the amendment causes a decrease in Cr, Ni, and Pb concentration in plants grown in both soils and a decrement in Ba and Co concentration in CA plants. On the contrary, the addition of AM induces an opposite trend of concentration for Cu and As in the aerial part, that is higher in the presence of the amendment both in CA and NOB soils.

Zinc, although an essential element for plants, has a behaviour similar to that assumed by most of the potentially toxic elements considered.

As observed for *L. sativa*, the concentration values of Pb and Cd in *B. oleracea* aerial part (Table S4) are below the limits set by Commission Regulation (EC) No. 1881/2006 of the Official Journal of the European Union [39], whether the improver is used or not.

The comparison of bioconcentration and translocation factors of *L. sativa* and *B. oleracea* grown in CA soil with or without AM, reported in Figures 4 and 5, and the results of ANOVA and Tukey tests reported in Tables S10–S13 of Supplementary Materials, provide evidence that the absorption of metals from the soil by *L. sativa* increases in the presence of an amendment for most elements, namely Cd, Co, Cr, Ni, Pb and Zn; on the contrary, a clear trend is not evidenced in the case of *B. oleracea*, since BF decreases or increase according to the considered element.

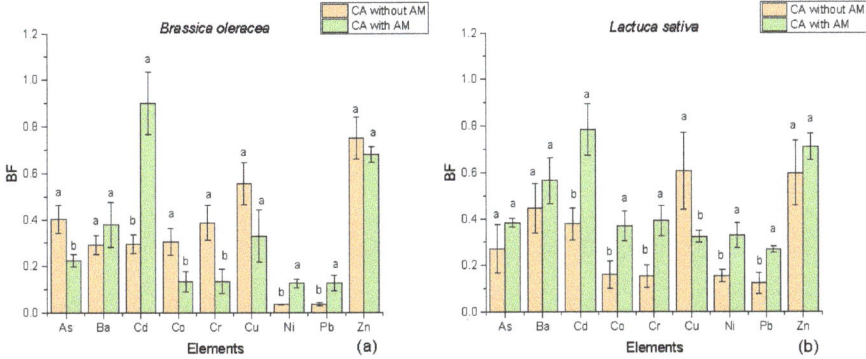

Figure 4. Bioconcentration factors (BF) of *Lactuca sativa* (**a**) and *Brassica oleracea* (**b**) grown in the polluted Ca soil with or without the amendment (AM). Letters above columns refer to the summary of multiple comparisons for pairs obtained with Tukey test. Letter **b** identifies BF values whose means are significantly lower than the ones obtained for the BF values identified by letter **a**.

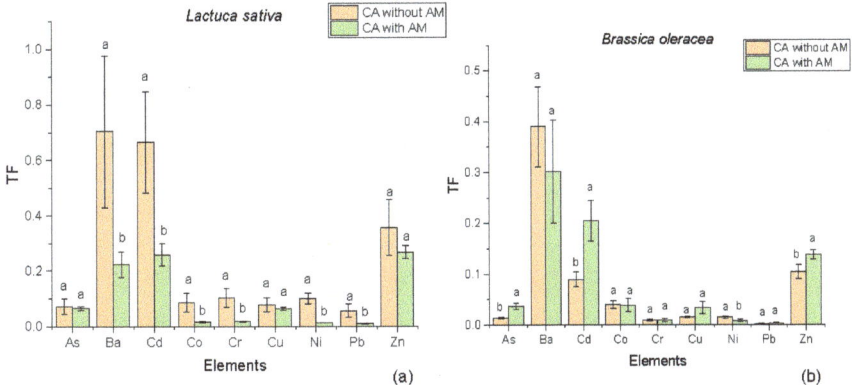

Figure 5. Translocation factors (TF) of *Lactuca sativa* (**a**) and *Brassica oleracea* (**b**) grown in the polluted CA soil with or without the amendment (AM). Letters above columns refer to the summary of multiple comparisons for pairs obtained with Tukey test. Letter **b** identifies TF values whose means are significantly lower than the ones obtained for the TF values identified by letter **a**.

Overall, in the presence of AM, *B. Oleracea* shows a lower absorption than *L. Sativa* for most elements examined.

Moreover, comparing the two vegetables, the translocation in the case of *B. oleracea* occurs to a much lesser extent than *L. sativa*. The translocation of potentially toxic elements from roots to the aerial part is significantly higher for *L. sativa* cultivated in the absence of an amendment, whereas no significant variation in TF values for *B. oleracea* were observed, with the exception of As, Cd and Zn.

Therefore, the addition of the amendment proved to be efficient in decreasing the translocation of potentially toxic elements in *L. sativa* and it did not alter the behavior of the *B. oleracea* which showed a lower assimilation of the toxic elements.

4. Conclusions

The site of Via Campana (Turin), usable as vegetable garden due to its central position, was found to be contaminated by the following metals: Cr, Zn, Ni, Pb and Co, whose concentrations exceeded Italian legislation limits.

The cultivation in growth chamber of the edible species *L. sativa* and *B. Oleracea* in the soil collected from CA, and the subsequent analysis of their aerial part and roots highlighted the absorption of the main potentially toxic elements by these plants. Comparing the PTEs uptake in *L. sativa* and *B. oleracea*, cultivated in CA soil, and in another soil considered as blank, it was evident that the metal content in vegetables was higher in the former than in the latter, as a consequence of the contamination of CA soil.

Despite this, the use of the soil amendment leads to a lower absorption with regard to *B. oleracea* and involves a considerable decrease in the translocation factor from the roots to the aerial part, especially for *L. sativa*, an encouraging result since it is precisely the aerial part that is consumed by humans, and therefore it is particularly important that there is a low content of contaminants in this portion.

The analyses showed that the presence of inorganic contaminants caused a reduction in the biomass of the plants, however, this can be compensated for by adding a soil amendment, thanks to the consequent enrichment of nutrients and organic matter, the improvement in the structure of the soil and the reduction in the bioavailability of toxic compounds. This is particularly convenient if we consider that the soil amendment also causes a decrease in the translocation of potentially toxic elements, thus not causing an increase in the concentration in the aerial parts that can be consumed.

Supplementary Materials: The following are available online at http://www.mdpi.com/2076-3417/10/13/4483/s1, Table S1. Summary of multiple comparisons for pairs of plants weights and morphometric parameters with Tukey test. NOB = plants grown in Nobile soil, CA = plants grown in Campana soil, AM = addition of amendment to the soil, Ls = *Lactuca sativa*, Bo = *Brassica oleracea*. Letter B identifies weights and parameters whose means are significantly lower than the ones obtained for weights and parameters identified by letter A; the same concept applies to letter C with respect to letter B, letter D with respect to letter C and letter E with respect to letter D, Table S2. Elements concentration [mg kg^{-1} dry weight] ± SD in *L. sativa* aerial part. NOB = plants grown in Nobile soil, CA = plants grown in Campana soil, AM = amendment, Table S3. Elements concentration [mg kg^{-1} dry weight] ± SD in *L. sativa* roots. NOB = plants grown in Nobile soil, CA = plants grown in Campana soil, AM = amendment, Table S4. Elements concentration [mg kg^{-1} dry weight] ± SD in *B. oleracea* aerial part. NOB = plants grown in Nobile soil, CA = plants grown in Campana soil, AM = amendment, Table S5. Elements concentration [mg kg^{-1} dry weight] ± SD in *B. oleracea* roots. NOB = plants grown in Nobile soil, CA = plants grown in Campana soil, AM = amendment, Table S6: Summary of multiple pairwise comparisons on element concentrations in *Lactuca sativa* roots obtained with ANOVA and Tukey tests. NOB = plants grown in Nobile soil, CA = plants grown in Campana soil, AM = addition of amendment to the soil. Letter B identifies soil samples whose mean concentrations in the respective elements are significantly lower than the ones obtained for soil samples identified by letter A; the same concept applies to letter C with respect to letter B, Table S7: Summary of multiple pairwise comparisons on element concentrations in *Lactuca sativa* aerial part obtained with ANOVA and Tukey tests. NOB = plants grown in Nobile soil, CA = plants grown in Campana soil, AM = addition of amendment to the soil. Letter B identifies plant samples whose mean concentrations in the respective elements are significantly lower than the ones obtained for plant samples identified by letter A; the same concept applies to letter C with respect to letter B and letter D with respect to letter C, Table S8: Summary of multiple pairwise comparisons on element concentrations in *Brassica oleracea* roots obtained with ANOVA and Tukey tests. NOB = plants grown in Nobile soil, CA = plants grown in Campana soil, AM = addition of amendment to the soil. Letter B identifies plant samples whose mean concentrations in the respective elements are significantly lower than the ones obtained for plant samples identified by letter A; the same concept applies to letter C with respect to letter B and letter D with respect to letter C, Table S9: Summary of multiple pairwise comparisons on element concentrations in *Brassica oleracea* aerial part obtained with ANOVA and Tukey tests. NOB = plants grown in Nobile soil, CA = plants grown in Campana soil, AM = addition of amendment to the soil. Letter B identifies plant samples whose mean concentrations in the respective elements are significantly lower than the ones obtained for plant samples identified by letter A; the same concept applies to letter C with respect to letter B and letter D with respect to letter C, Table S10: Summary of multiple comparisons for pairs of *Lactuca sativa* BF with Tukey test. CA = plants grown in Campana soil, AM = addition of amendment to the soil. Letter B identifies BF values whose means are significantly lower than the ones obtained for BF values identified by letter A, Table S11: Summary of multiple comparisons for pairs of *Lactuca sativa* TF with Tukey test. CA = plants grown in Campana soil, AM = addition of amendment to the soil. Letter B identifies TF values whose means are significantly lower than the ones obtained for TF values identified by letter A, Table S12: Summary of multiple comparisons for pairs of *Brassica oleracea* BF with Tukey test. CA = plants grown in Campana soil, AM = addition of amendment to the soil. Letter B identifies BF values whose means are significantly lower than the ones obtained for BF values identified by letter A, Table S13: Summary of multiple comparisons for pairs of *Brassica oleracea* TF with Tukey test. CA = plants grown in Campana soil, AM = addition of amendment to the soil. Letter B identifies TF values whose means are significantly lower than the ones obtained for TF values identified by letter A.

Author Contributions: Conceptualization, A.F. and P.C.; methodology, A.F. and M.M. (Marco Mucciarelli); formal analysis, E.G.; investigation, D.F. and G.B.; resources, M.M. (Mery Malandrino), D.F. and A.F.; data curation, G.B.; writing—original draft preparation, E.G. and D.F.; writing—review and editing, M.M. (Marco Mucciarelli), M.M. (Mery Malandrino), O.A.; visualization, E.G.; supervision, M.M. (Mery Malandrino), A.F. and P.C.; project administration, M.M. (Mery Malandrino) and P.C.; funding acquisition, P.C. All authors have read and agreed to the published version of the manuscript.

Funding: This work has received funding from Compagnia di San Paolo (project CSTO168877- ReHorti).

Acknowledgments: We kindly acknowledge ACEA Pinerolese for providing the soil improver.

Conflicts of Interest: The authors declare no conflict of interest.

References

1. C.d.T.-. Orti urbani. Available online: www.comune.torino.it (accessed on 9 February 2020).
2. DeLind, L.B. Place, work, and civic agriculture: Common fields for cultivation. *Agr. Hum. Values* **2002**, *19*, 217–224. [CrossRef]
3. Orsini, F.; Kahane, R.; Nono-Womdim, R.; Gianquinto, G. Urban agriculture in the developing world: a review. *Agron. Sustain. Dev.* **2013**, *33*, 695–720. [CrossRef]
4. Leake, J.R.; Adam-Bradford, A.; Rigby, J.E. Health benefits of 'grow your own' food in urban areas: implications for contaminated land risk assessment and risk management? *Environ. Health* **2009**, *8*, S6. [CrossRef] [PubMed]
5. La Greca, P.; La Rosa, D.; Martinico, F.; Privitera, R. Agricultural and green infrastructures: the role of non-urbanised areas for eco-sustainable planning in a metropolitan region. *Environ. Pollut.* **2011**, *159*, 2193–2202. [CrossRef]
6. Muganu, M.; Balestra, G.M.; Senni, S. THE IMPORTANCE OF ORGANIC METHOD IN SOCIAL HORTICULTURE. In *ISHS Acta Horticulturae 881: II International Conference on Landscape and Urban Horticulture*; ISHS: Leuven, Belgium, 2010; pp. 847–849.
7. Comune di Torino. Available online: http://www.comune.torino.it/verdepubblico/patrimonioverde/verdeto/numeri.shtml (accessed on 2 February 2020).
8. Antisari, L.V.; Orsini, F.; Marchetti, L.; Vianello, G.; Gianquinto, G. Heavy metal accumulation in vegetables grown in urban gardens. *Agro. Sustain. Dev.* **2015**, *35*, 1139–1147. [CrossRef]
9. Alloway, B.J. Contamination of soils in domestic gardens and allotments: a brief overview. *Land Contam. Reclam.* **2004**, *12*, 179–187. [CrossRef]
10. Hough, R.L.; Breward, N.; Young, S.D.; Crout, N.M.; Tye, A.M.; Moir, A.M.; Thornton, I. Assessing potential risk of heavy metal exposure from consumption of home-produced vegetables by urban populations. *Environ. Health Perspect.* **2004**, *112*, 215–221. [CrossRef] [PubMed]
11. McBride, M.B.; Simon, T.; Tam, G.; Wharton, S. Lead and Arsenic Uptake by Leafy Vegetables Grown on Contaminated Soils: Effects of Mineral and Organic Amendments. *Water Air Soil Pollut.* **2013**, *224*. [CrossRef]
12. Kaur, H.; Garg, N. Recent Perspectives on Cross Talk Between Cadmium, Zinc, and Arbuscular Mycorrhizal Fungi in Plants. *J. Plant Growth Reg.* **2017**, 1–14. [CrossRef]
13. Khan, A.; Khan, S.; Khan, M.; Qamar, Z.; Waqas, M. The uptake and bioaccumulation of heavy metals by food plants, their effects on plants nutrients, and associated health risk: A review. *Environ. Sci. Pollut. Res. Int.* **2015**, *22*. [CrossRef]

14. Clemens, S. Toxic metal accumulation, responses to exposure and mechanisms of tolerance in plants. *Biochimie* **2006**, *88*, 1707–1719. [CrossRef] [PubMed]
15. Gaggero, E.; Malandrino, M.; Fabbri, D.; Bordiglia, G.; Fusconi, A.; Mucciarelli, M.; Inaudi, P.; Calza, P. Uptake of Potentially Toxic Elements by Four Plant Species Suitable for Phytoremediation of Turin Urban Soils. *Appl. Sci.* **2020**, *10*, 3948. [CrossRef]
16. ImageJ. Available online: https://imagej.net/ImageJ (accessed on 6 April 2020).
17. *Metodi Ufficiali di Analisi Chimica del Suolo*; Decreto Ministeriale, Ministero Delle Politiche Agricole e Forestali: Rome, Italy, 13 September 1999.
18. Schulte, E.; Hoskins, B. Recommended Soil Testing Procedures for the Northeastern United States. *Recomm. Soil Org. Matter Tests* **1995**, *493*, 52–60.
19. Thompson, M.; Ramsey, M.H.; Coles, B.J. Communication. Interactive matrix matching: a new method of correcting interference effects in inductively coupled plasma spectrometry. *Analyst* **1982**, *107*, 1286–1288. [CrossRef]
20. Yoon, J.; Cao, X.; Zhou, Q.; Ma, L.Q. Accumulation of Pb, Cu, and Zn in native plants growing on a contaminated Florida site. *Sci. Total Environ.* **2006**, *368*, 456–464. [CrossRef]
21. Fitz, W.J.; Wenzel, W.W. Arsenic transformations in the soil rhizosphere plant system fundamentals and potential application to phytoremediation. *J. Biotechnol.* **2002**, *99*, 259–278. [CrossRef]
22. Einax, W.; Zwanziger, H.W.; Gei, S. *Chemometrics in Environmental Analysis*; Wiley-VHC: Weinhem, Germany, 1997.
23. Massart, D.L.; Vandenginste, B.G.M.; Buydens, L.M.C.; De Jono, S.; Leqi, P.J.; Smeyers-Verbeke, J. *Handbook of Chemometrics and Quantimetrics, Parts A and B*; Elsevier: Amsterdam, The Netherlands, 1997.
24. Lorenz, K.; Lal, R. The Depth Distribution of Soil Organic Carbon in Relation to Land Use and Management and the Potential of Carbon Sequestration in Subsoil Horizons. *Adv. Agron.* **2005**, *88*, 35–66.
25. ARPAV. Available online: https://www.arpa.veneto.it/arpavinforma/indicatori-ambientali/indicatori_ambientali/geosfera/qualita-dei-suoli/contenuto-di-carbonio-organico-nello-strato-superficiale-di-suolo/view (accessed on 14 February 2020).
26. Dube, A.; Zbytniewski, R.; Kowalkowski, T.; Cukrowska, E.; Buszewski, B. Adsorption and migration of heavy metals. *Pol. J. Environ. Stud.* **2001**, *10*, 1–10.
27. McCauley, A.; Jones, C.; Jacobsen, J. Soil pH and organic matter. *Nutr. Manag.* **2009**, *8*, 1–12.
28. Yin, Y.; Impellitteri, C.A.; You, S.-J.; Allen, H.E. The importance of organic matter distribution and extract soil:solution ratio on the desorption of heavy metals from soils. *Sci. Total Environ.* **2002**, *287*, 107–119. [CrossRef]
29. Lair, G.J.; Gerzabek, M.H.; Haberhauer, G. Sorption of heavy metals on organic and inorganic soil constituents. *Environ. Chem. Lett.* **2007**, *5*, 23–27. [CrossRef]
30. *Norme in Materia Ambientale*. Decreto Legislativo, Italy, 3 April 2006; n. 152. Available online: https://www.ambientediritto.it/Legislazione/VARIE/2006/dlgs_2006_n.152.htm (accessed on 24 June 2020).
31. Syers, J.; Johnston, A.; Curtin, D. Efficiency of soil and fertilizer phosphorus use. *FAO Fertil. Plant Nutr. Bull.* **2008**, *18*.
32. McBride, M.B.; Shayler, H.A.; Spliethoff, H.M.; Mitchell, R.G.; Marquez-Bravo, L.G.; Ferenz, G.S.; Russell-Anelli, J.M.; Casey, L.; Bachman, S. Concentrations of lead, cadmium and barium in urban garden-grown vegetables: The impact of soil variables. *Environ. Pollut.* **2014**, *194*, 254–261. [CrossRef] [PubMed]
33. Ross, S.M. *Toxic Metals in Soil-plant Systems*; Wiley: Hoboken, NJ, USA, 1994; p. 484.
34. Tong, S.; von Schrinding, Y.E.; Prapamontol, T. Environmental lead exposure: a public health problem of global dimension. *Bull. World Health Organ.* **2000**, *78*, 1068–1077. [PubMed]
35. Kushwaha, A.; Hans, N.; Kumar, S.; Rani, R. A critical review on speciation, mobilization and toxicity of lead in soil-microbe-plant system and bioremediation strategies. *Ecotoxicol. Environ. Saf.* **2018**, *147*, 1035–1045. [CrossRef]
36. Wuana, R.A.; Okieimen, F.E. Heavy Metals in Contaminated Soils: A Review of Sources, Chemistry, Risks and Best Available Strategies for Remediation. *ISRN Ecol.* **2011**, *2011*, 1–20. [CrossRef]
37. Su, C. A review on heavy metal contamination in the soil worldwide: Situation, impact and remediation techniques. *Environ. Skept. Crit.* **2014**, *3*, 24.

38. Tchounwou, P.B.; Yedjou, C.G.; Patlolla, A.K.; Sutton, D.J. Heavy metal toxicity and the environment. *Exp. Suppl.* **2012**, *101*, 133–164.
39. Biasioli, M.; Barberis, R.; Ajmone-Marsan, F. The influence of a large city on some soil properties and metals content. *Sci. Total Environ.* **2006**, *356*, 154–164. [CrossRef]
40. Bonifacio, E.; Falsone, G.; Piazza, S. Linking Ni and Cr concentrations to soil mineralogy: Does it help to assess metal contamination when the natural background is high? *Soils Sediments* **2010**, *10*. [CrossRef]
41. Commission Regulation (EC) No. 1881/2006 of 19 December 2006 setting maximum levels for certain contaminants in foodstuffs. *Off. J. Eur. Union* **2006**, *364*, 324–365.

© 2020 by the authors. Licensee MDPI, Basel, Switzerland. This article is an open access article distributed under the terms and conditions of the Creative Commons Attribution (CC BY) license (http://creativecommons.org/licenses/by/4.0/).

Article

A Study on the Flow Characteristics of Copper Heavy Metal Microfluidics with Hydrophobic Coating and pH Change

Jung-Geun Han [1,2], Dongho Jung [3], Jong-Young Lee [1], Dongchan Kim [4,*] and Gigwon Hong [5,*]

1. School of Civil and Environmental Engineering, Urban Design and Study, Chung-Ang University, Seoul 06974, Korea; jghan@cau.ac.kr (J.-G.H.); geoljy@cau.ac.kr (J.-Y.L.)
2. Department of Intelligent Energy and Industry, Chung-Ang University, Seoul 06974, Korea
3. Department of Civil Engineering, Chung-Ang University, Seoul 06974, Korea; curara81@gmail.com
4. Department of Infrastructure Safety Research, Korea Institute of Civil Engineering and Buldign Technology, Goyang-si 10223, Korea
5. Department of Civil and Disaster Prevention Engineering, Halla University, Wonju-si 26404, Korea
* Correspondence: dc_kim@kict.re.kr (D.K.); g.hong@halla.ac.kr (G.H.)

Abstract: The present study purpose was to identify the flow characteristics of the drainage filter considering the characteristics of the landfill site, and to study the basic technology for efficient remediation of heavy metals. To this end, copper heavy metal was selected in consideration of landfill characteristics, and a study on flow characteristics was conducted using hydrophobic coated capillary tubes and microparticles. It was confirmed that the flow rate decreased as the pH increased at the hydrophobic surface, and pH 4, 6, and 8 flowed similarly in the center of the capillary tube, but decreased at pH 10. In the bottom part, it moved at the slowest speed of 1~4 μm/s and middle of center moved 17~25 μm/s. There was little change in flow in the CFD (Computational Fluid Dynamics) numerical analysis considering the surface contact angle, which is a hydrophobic characteristic, and the velocity coefficient was presented by regression analysis through the experimental results. In this way, the current study will be a basic examination of the selective remediation process of pH on hydrophobic coated surfaces.

Keywords: hydrophobic; copper; microfluidics; pH change; remediation

1. Introduction

Recently, the importance of soil environment has been recognized, primarily because the seriousness of soil and groundwater pollution has been heightened due to waste and harmful chemicals generated by population growth and industrial development worldwide [1–3]. Pollution of the ground and groundwater by pollutants occurs through various routes. If pollutants generated from landfill leachate, factory wastewater, pesticides, and abandoned mines are discharged without going through the treatment process, it can cause serious problems [4–6]. In addition, soil has a very large buffering capacity for pollutants, but its limited capacity varies depending on the characteristics of the soil and environmental factors [7]. In particular, soil pollution caused by chemical spills from chemical substance handling facilities is difficult to manage efficiently and adversely affects human health compared to water and air pollution [8–11].

On the other hand, the technology for removing and restoring pollutants in the soil has been developed and is practical in application. Among them, soil cleaning, soil steam extraction, water treatment, and bio-venting are mainly used as field restoration technologies in contaminated areas [12–14]. However, the above restoration techniques are mainly applied to coarse and unsaturated ground, and their applicability and efficiency are inevitably reduced to grounds with low permeability due to the large amount of fines contained in areas such as landfill industrial complexes. Further, in the past landfilling process, various household wastes were buried, so a detailed investigation and soil management in these areas is urgently required [15–19].

As an example of a remediation technology that can solve this problem, an example is a remediation technology using drainage materials consisting of filters and cores, which has been under a lot of research recently. This has the disadvantage of reducing the permeability of the drainage filter material due to various chemical reactions caused by the composition of complex pollutants in the soil, which can lead to inefficient results in decontamination. Although treatment techniques are needed to satisfy the flow characteristics of pollutants in the soil, there is little research on the flow characteristics of pollutants at this time. Therefore, it is necessary to identify the flow characteristics of complex pollutants contained in soil and groundwater when passing through the transmission membrane in the filter to remove pollutants, and development to purify complex pollutants based on them.

The study aimed to evaluate the application of the separation of various particles such as biomolecules and gas filters to the remediation of contaminants in the nanoporous permeable membrane, which is applied as a filter material [20,21]. Based on the fact that there are permeability coefficients which are the main variables of pollutant flow and are different because the ground is composed of various types of strata and soil, it is necessary to study the flow characteristics of heavy metal complex pollutants for complex pollutant flow control.

In this study, most industrial complexes are built in landfills, and the main purpose is to study soil pollution remediation techniques considering the location characteristics of landfills where various heavy metal complex pollutants exist. However, little research has been done on selective remediation in contaminated ground at the same time as improving soft ground using drainage materials. The purpose of this study is to investigate the flow characteristics of high concentration heavy metal contaminated soil and groundwater passing through the permeable membrane in the filter for removing contaminants. This paper is composed of the materials used in the experiment, the method, and the conditions of CFD analysis. In addition, the tube for simulating the pores passing through the drainage filter was subjected to hydrophobic treatment, and the moving velocity of copper contaminants was confirmed through microparticles. This paper was composed by drawing conclusions based on the experimental results.

2. Materials and Methods

2.1. Materials

Industrial complexes, such as landfills, were contaminated with a variety of heavy metals, and copper was chosen to account for a large portion of them. As heavy metal copper contaminants, copper sulfate $Cu(SO_4)$ was dissolved in distilled water and a silage pump was used for constant injection. Capillary tubes were used to simulate drainage filters, and optical microscopes were used to monitor the movement of microparticles. Microparticles were used to check the flow rate on the Capillary tube. The capillary tube used in the study was square, the inner diameter was 400 µm, and an optical microscope (Keyence, Osaka, Japan, VHX-1000) capable of 5000 times magnification was used. Polystyrene microparticles having a size of 2 µm were used. Microparticles have 5.4×10^9 particles per 1 mL, with a weight of 1.05 g/cm^3.

2.2. Methods

2.2.1. Microfluidic Flow Test

Contaminants were made based on the soil pollution standard of 6000 ppm indicated in the Environmental White Paper [22] for heavy copper metal contaminants, and the pH was divided into 4, 6, 8, and 10 for acid and basic conditions. The inflow velocity was injected consistently using a silage pump. The inflow velocity into the capillary tube was applied 1.0×10^{-3} cm/s considering the pH conditions, and 0.1 M of HCl and 0.1 M of NaOH were used for pH control. The conditions for each experiment are shown in Table 1, and the experimental equipment and microparticles are shown in Figures 1 and 2. The tube was injected with a mixture of copper and particles, and the flow velocity was measured. A square capillary tube was used to analyze flow characteristics, and a square capillary tube was observed in half as shown in Figure 3. Since the square capillary tube was used, it was symmetrical up and down. The half-section area was observed, and the flow of microfluids was measured using microparticles by observing 3 parts for the bottom and center of the side and central part of the capillary tube. Three microparticles were selected using a video taken with an optical microscope. Using the video program, the moving distance of the microparticles was calculated after a certain period of time.

Table 1. Microfluidic Flow Test Condition.

Classification	Test Condition			
	Pollutants	Concentration (ppm)	Flow Velocity (cm/s)	pH
Flow characteristic of pollutants	Copper	6000	1.0×10^{-3}	4, 6, 8, 10

Figure 1. Test equipment.

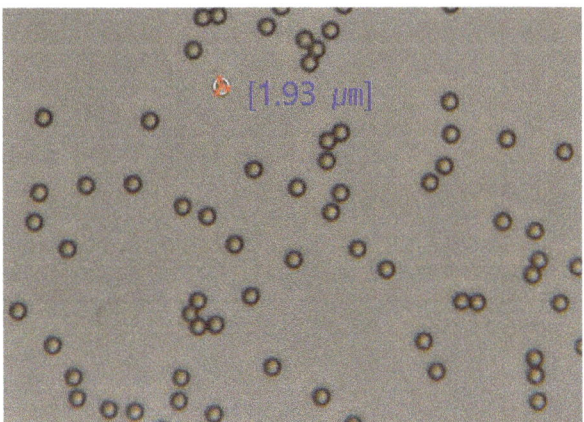

Figure 2. Microparticles (×3000).

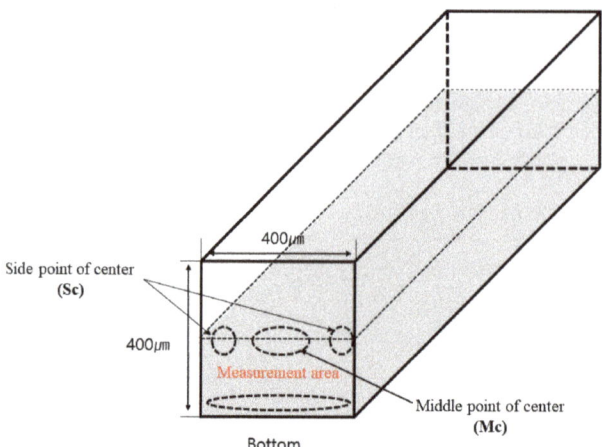

Figure 3. Square capillary tube schematic and observation position.

2.2.2. CFD Numerical Analysis

To check the microfluidic flow velocity in the square capillary tube according to the pH of heavy metals, CFD numerical analysis was conducted using Ansys. After hydrophobic treatment of the square capillary tube, the experimental value of the contact angle was applied to confirm the hydrophobic state, and the experimental value according to the pH was applied as the contact angle. In addition, glass-made square capillary tubes were applied for this microfluidic flow experiment, and heavy metal copper contaminants of 6000 ppm were applied. The properties used in the numerical analysis are shown in Table 2 below. Microfluidic flow test values applied with hydrophobic coating were applied.

Table 2. CFD properties.

Flow Rate (cm/s)	Contract Angle			
	pH 4	pH 6	pH 8	pH 10
1.0×10^{-3}	124°	125°	119°	116°

Capillary tube has a hydrophilic surface property of glass material, and the solvent used to prepare heavy metal solutions is also distilled water and is a hydrophilic aqueous solution. The hydrophobic coating was applied with an organic solution, and the hydrophobic coating was used to check the flow characteristics of the hydrophobic surface in the capillary tube. The contact angle was confirmed for heavy metal contaminants according to distilled water and pH conditions. The surface contact angle—when passing through heavy metal contaminants after hydrophobic coating—produced some interesting results, such as the contact angle becoming decreased with basicity. Table 2 below shows the contact angle of the capillary tube identified through the previous study and the properties applied to the numerical analysis.

In order to confirm the flow of microfluids in the hydrophobic coated capillary tube, the numerical analysis model was modeled on the square capillary tube. The following Figure 4 is a modeling of a square capillary tube, which is modeled with an inner diameter of 400 μm and an outer diameter of 800 μm and a length of 3 cm.

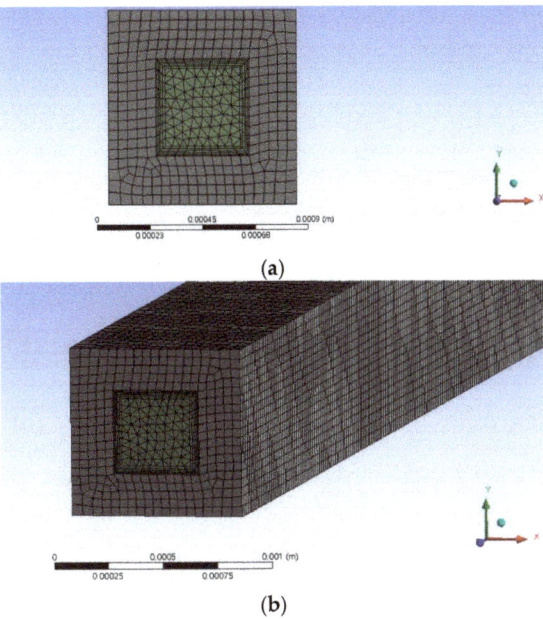

Figure 4. Capillary tube mesh: (**a**) Section 1; (**b**) Section 2.

3. Results and Analysis

3.1. Microfluidic Flow Characteristic

This test observed microfluidic flow characteristics for heavy metal copper contaminants.

3.1.1. pH 4

Figure 5 is the result of observation of the flow characteristics of copper contaminants under the condition of pH 4. After the start of the experiment, movement over time was observed based on the microparticles (red circle) at the far right of the screen. As a result of observing the moving distance by selecting three microparticles, it was confirmed that the microparticles moved to an average of 4 μm/s. Figures 6 and 7 represent the moving distances of the side and center at the middle position in the capillary tube, and it was confirmed that the side surface was about 11 μm/s and the center was about 25 μm/s. In other words, it was found that the central center of the capillary tube moved rapidly.

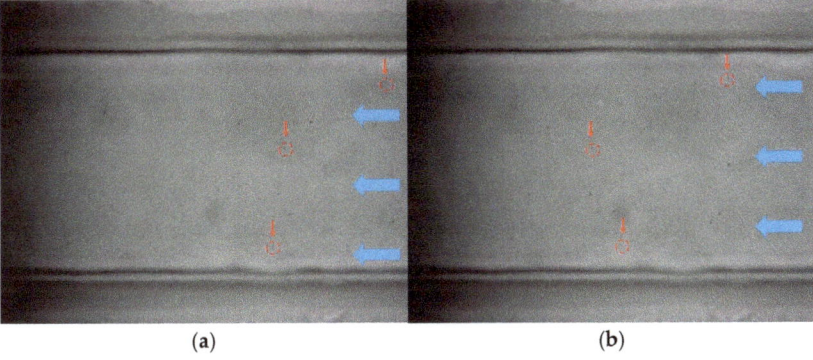

Figure 5. Experimental results of flow characteristics of copper contaminants–capillary tube bottom: (**a**) particle position before moving; (**b**) particle position after moving.

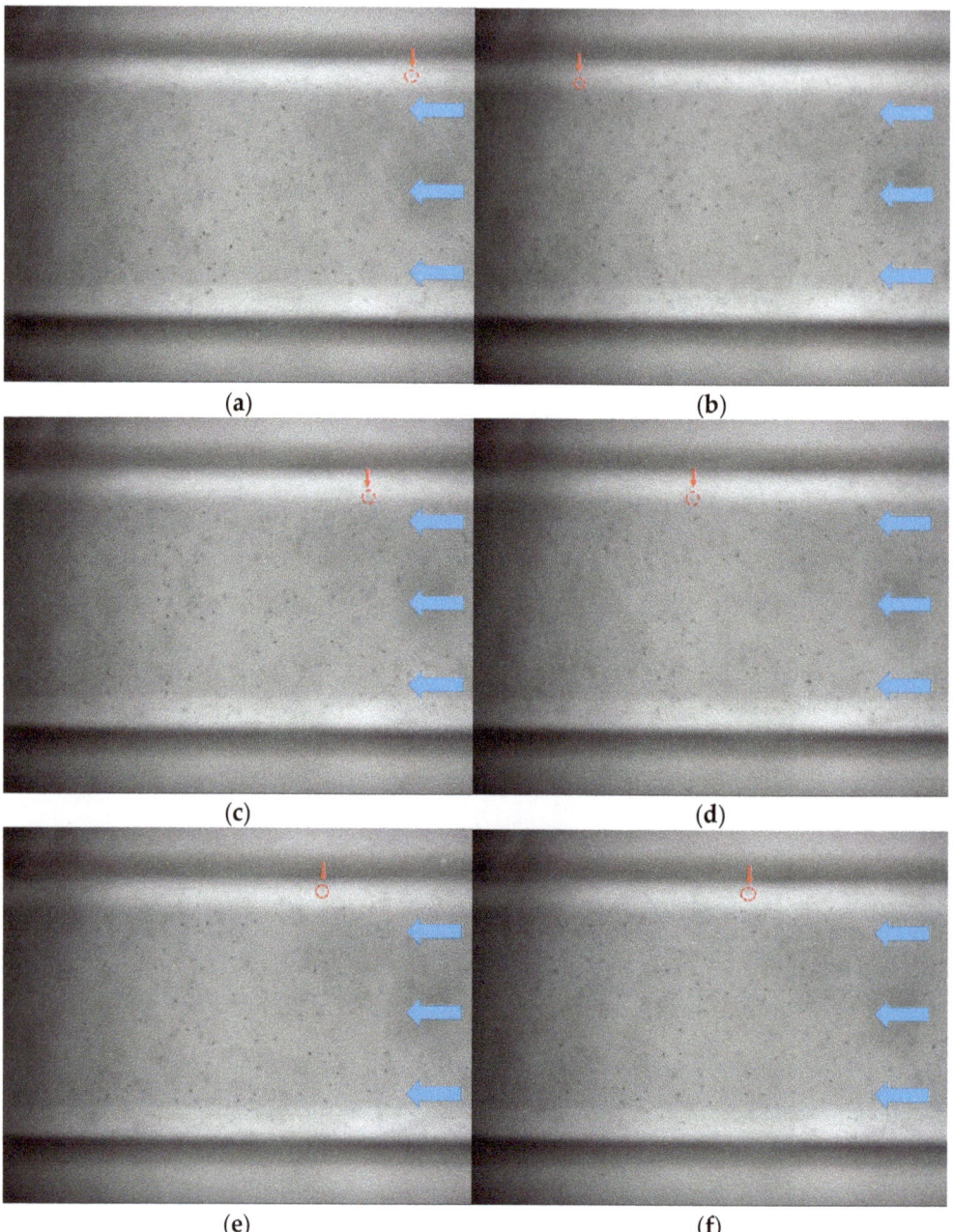

Figure 6. Experimental results of flow characteristics of copper contaminants–capillary tube side of center: (**a**) Particle position before moving 1; (**b**) particle position after moving 1; (**c**) particle position before moving 2; (**d**) particle position after moving 2; (**e**) particle position before moving 3; (**f**) particle position after moving 3.

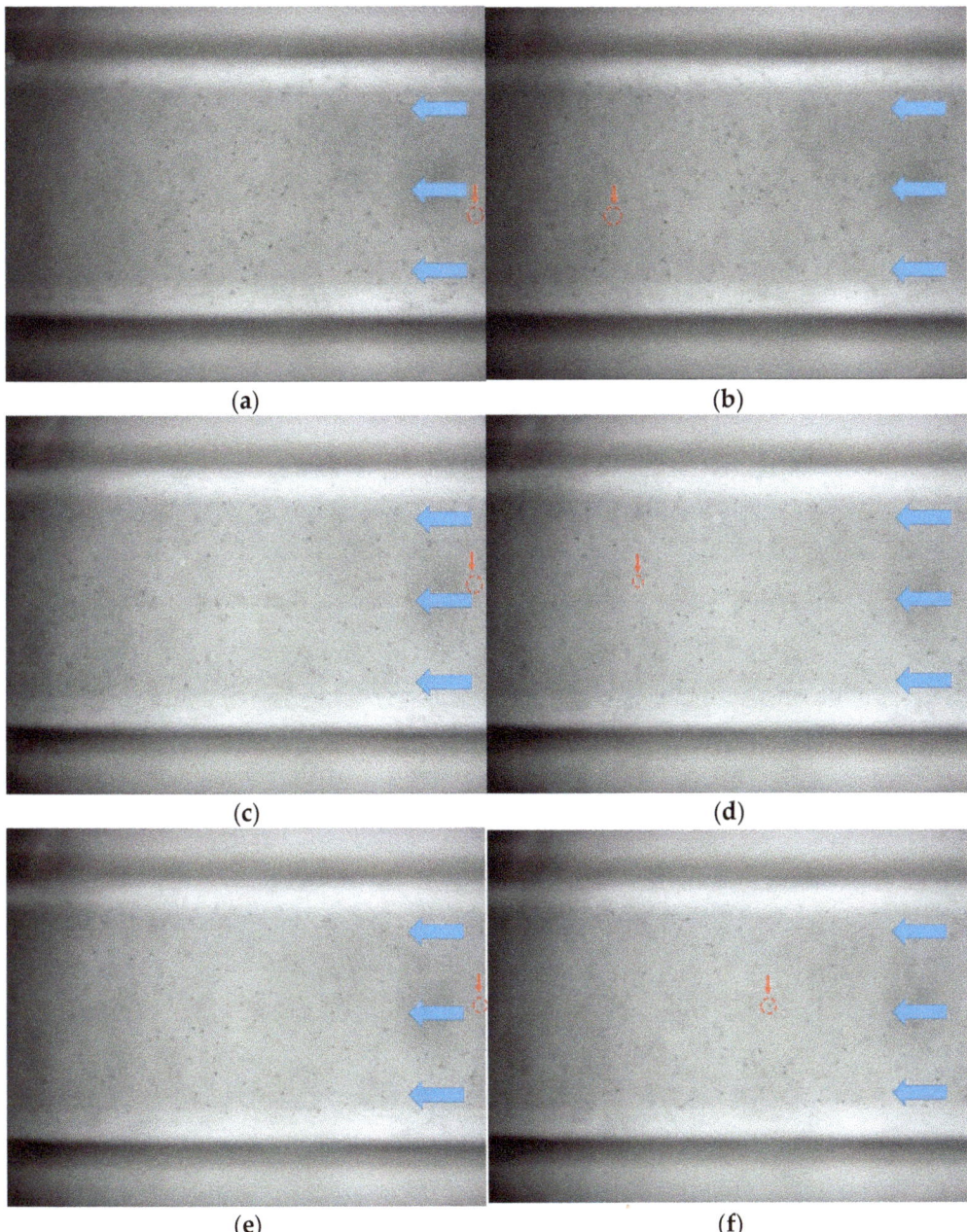

Figure 7. Experimental results of flow characteristics of copper contaminants–capillary tube middle of center: (**a**) Particle position before moving 1; (**b**) particle position after moving 1; (**c**) particle position before moving 2; (**d**) particle position after moving 2; (**e**) particle position before moving 3; (**f**) particle position after moving 3.

3.1.2. pH 6

The flow velocity was measured in the same way as pH 4 in Section 3.1.1. As a result of observing the movement of three particles over time after the start of the experiment, it was found that the movement distance of the microparticles slightly decreased compared to the case of pH 4. In the capillary tube side of center and middle of center, the travel distance of the central side of the capillary tube is about 10 μm/s and the center is about 21 μm/s, which is slower than pH 4. In addition, the capillary tube bottom confirmed that the movement distance of microparticles at the bottom was about 3 μm/s.

3.1.3. pH 8

The flow velocity was measured in the same way as pH 4 in Section 3.1.1. The moving distance of the microparticles at the bottom was confirmed to be about 2 μm/s, and the side and center of the central part in the capillary tube were about 11 μm/s and about 25 μm/s.

3.1.4. pH 10

The flow velocity was measured in the same way as pH 4 in Section 3.1.1. The moving distance of microparticles at the bottom was confirmed to be about 1 μm/s, and the side and center of the central part in the capillary tube were about 9 μm/s and about 17 μm/s. These results indicate a tendency for the distance traveled to decrease as pH increases.

3.2. Microfluid Flow Moving Distance

The following Figure 8 shows the moving distance over time according to pH for microparticles on the capillary tube. In the bottom part, it moved at the slowest speed of 1~4 μm/s, and it was confirmed that the influence on the microfluid flow was small at the rest of the pH except for pH 10 in the center part. There was little effect at pH 4–8, and it was confirmed that the flow rate decreased at pH 10. As a result, it is confirmed that if the hydrophobic coating is applied under high pH conditions, the flow rate of heavy metal contaminants can be controlled.

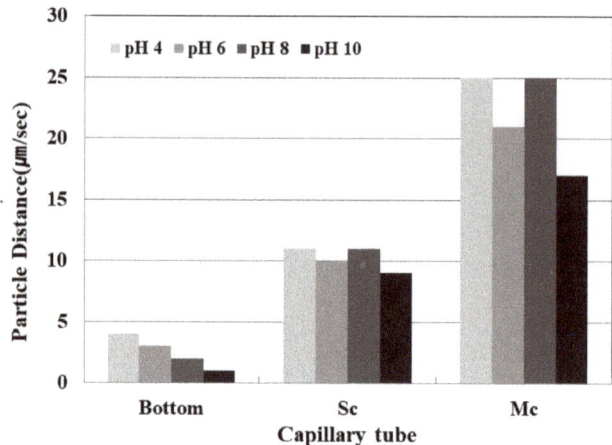

Figure 8. Micro particles distance (μm/s).

3.3. Microparticle Flow Velocity

The results of the analysis of flow characteristics for the bottom and central sides/centers of Capillary tubes for copper contaminants are shown in Figure 9. The flow rate at the bottom of the capillary tube was about 1.0×10^{-4} cm/s, which is slower than the inflow rate of 1.0×10^{-3} cm/s, and the flow rate at the center was analyzed to be 1.0×10^{-3} cm/s similar to the inflow rate. In addition, it was analyzed that the flow rate from the center to the center appeared somewhat faster than the side. When copper contaminants pass through the filter, it was revealed that the inflow of contaminants could be controlled depending on whether the surface's hydrophobic properties persist.

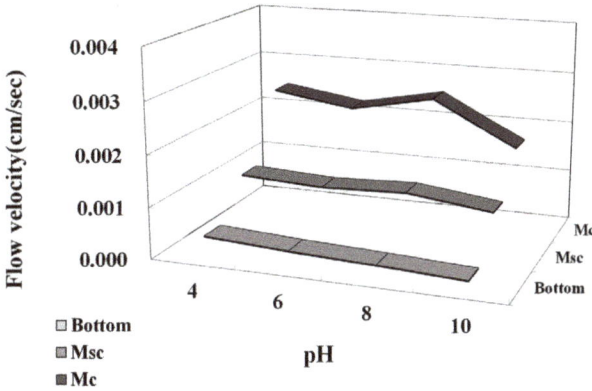

Figure 9. Microparticle flow velocity (μm/s).

3.4. CFD Numerical Analysis

In the hydrophobic coated square capillary tube, the microfluidic flow according to the pH did not change at pH 4, pH 6 and pH 8, and it was confirmed that the speed decreased slightly at pH 10. Table 3 shows the flow rate at the center, which is the fastest among the speeds in the capillary tube among numerical analysis results, and it is possible to confirm the flow rate slower than the experimental results. The following Figures 10–13 show the result of numerical analysis of flow velocity and vector of square capillary tube through numerical analysis.

Table 3. CFD analysis result.

	pH			
	4	6	8	10
Flow rate (cm/s)	1.272×10^{-3}	1.272×10^{-3}	1.272×10^{-3}	1.264×10^{-3}

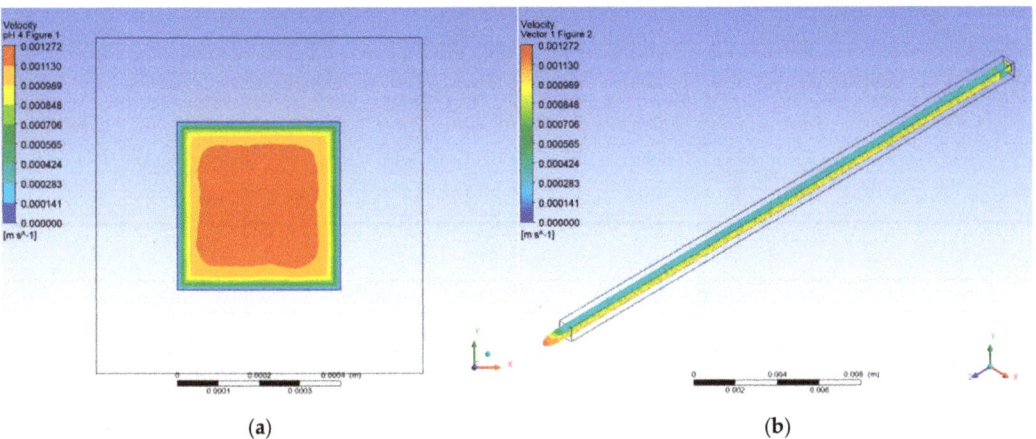

Figure 10. CFD numerical analysis results (pH 4): (a) contour; (b) vector.

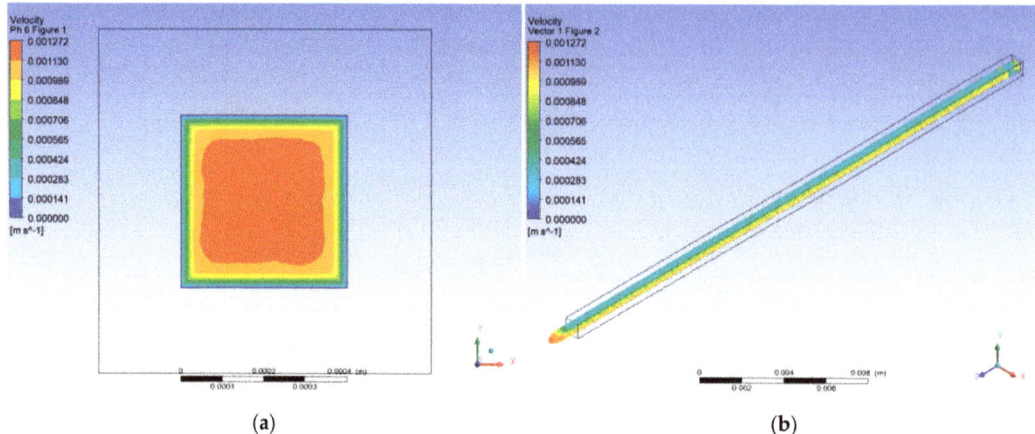

Figure 11. CFD numerical analysis results (pH 6): (**a**) contour; (**b**) vector.

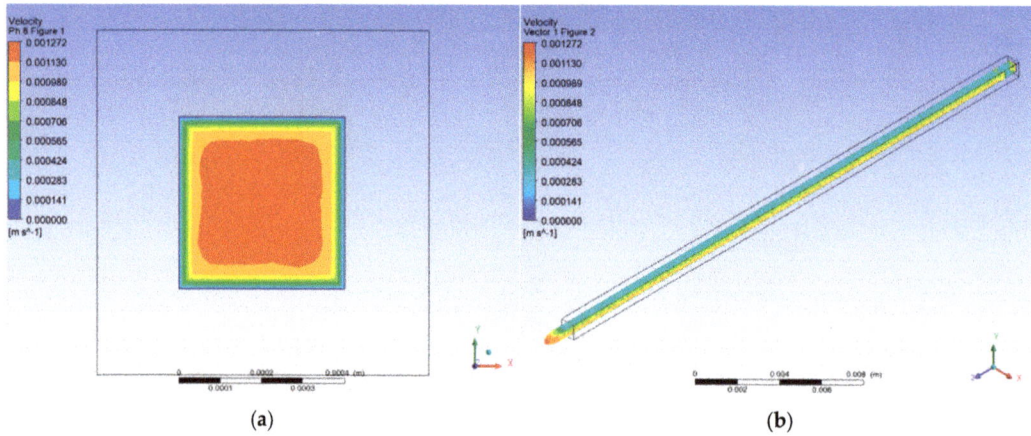

Figure 12. CFD numerical analysis results (pH 8): (**a**) contour; (**b**) vector.

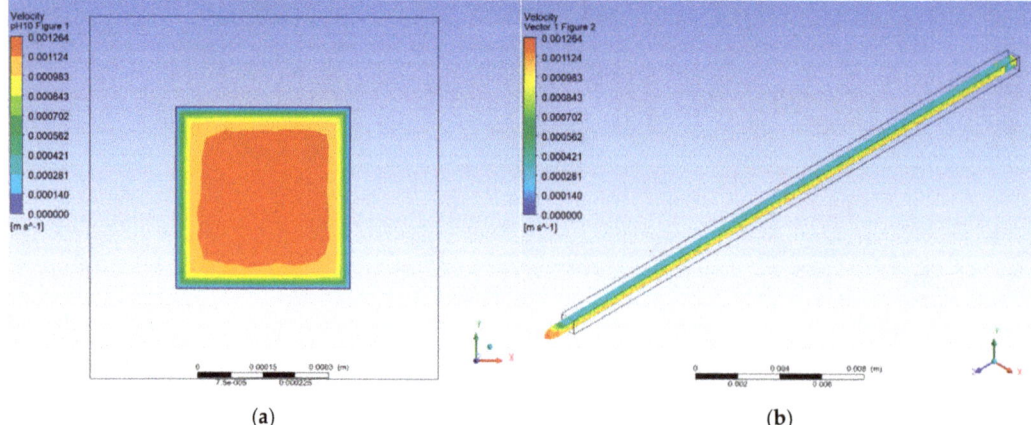

Figure 13. CFD numerical analysis results (pH 10): (**a**) contour; (**b**) vector.

The numerical analysis shows that there was no change in velocity at pH 4, pH 6 and pH 8, and that there was a small decrease in velocity at pH 10, but the difference between the actual flow rate and the actual flow rate was confirmed to be caused by pH. Based on the experimental results, the velocity coefficient was presented. At this time, the diffusion coefficient was calculated by regression analysis of the flow rate measured for the copper contaminant and expressed in the rate coefficient Equations 1 to 3 according to the pH.

$$v_b = 0.0005e^{-0.335pH} \tag{1}$$

$$v_{sc} = 0.0013e^{-0.115pH} \tag{2}$$

$$v_{mc} = 0.0031e^{-0.192pH} \tag{3}$$

Here, v_b: velocity coefficient at the bottom of the capillary tube, v_{sc}: velocity coefficient at the central side of the capillary tube, v_{mc}: speed factor at the center of the capillary tube, and pH: pH of heavy metal contaminants are shown.

4. Conclusions

In this study, a microfluidic flow experiment was performed to apply a hydrophilic hydrophobic filter for selective remediation of heavy metal copper contaminated ground. Accordingly, the summary of the results of this study is as follows.

(1) As a result of microfluidic flow experiments for heavy copper metal contaminants, it was confirmed that the movement of contaminants was small in locations where the influence of hydrophobic properties was large. This means that hydrophobic surfaces can slow down the flow of hydrophilic heavy metal contaminants.

(2) The analysis of the flow rate according to each experimental condition confirmed that the change in flow rate of pH conditions was an influential factor. It was confirmed that the flow rate increased as the pH was more basic, and it was confirmed that the hydrophobic properties slightly decreased under basic conditions.

(3) Numerical analysis was performed by applying the contact angle, which is a hydrophobic characteristic, however, it was difficult to compare and analyze through CFD analysis. Based on the flow rate according to pH as a result of the experiment, the velocity coefficient was proposed through regression analysis. Using the velocity coefficient, it is determined that the flow rate according to pH on a hydrophobic coated surface can be determined.

In this study, it was confirmed that heavy metal contaminants were slowed down through hydrophobic coating. Through additional laboratory and filed tests, selective remediation is possible in the ground contaminated with heavy metals and organic.

Author Contributions: Conceptualization, J.-G.H. and G.H.; methodology, J.-G.H. and G.H.; validation, J.-G.H., J.-Y.L. and G.H.; formal analysis, D.J. and D.K.; investigation, D.J., J.-Y.L. and D.K.; resources, J.-G.H., J.-Y.L. and G.H.; data curation, D.J. and D.K.; writing—original draft preparation, D.J. and D.K.; writing—review and editing, J.-G.H., D.J., J.-Y.L., D.K. and G.H.; visualization, J.-G.H., D.J., D.K. and G.H.; supervision, J.-G.H., J.-Y.L. and G.H.; project administration, D.J. and D.K. All authors have read and agreed to the published version of the manuscript.

Funding: This research was supported by the Chung-Ang University Research Scholarship Grants in 2011 and a grant from the National Research Foundation (NRF) of Korea, funded by the Korea government (MSIP) (NRF-2019R1A2C2088962).

Institutional Review Board Statement: Not applicable.

Informed Consent Statement: Not applicable.

Data Availability Statement: Data presented in this study are available on request from the corresponding author. The data are not publicly available due to data that are also part of an ongoing study.

Acknowledgments: This research was supported by the Chung-Ang University Research Scholarship Grants in 2011 and a grant from the National Research Foundation (NRF) of Korea, funded by the Korea government (MSIP) (NRF-2019R1A2C2088962).

Conflicts of Interest: The authors declare no conflict of interest.

References

1. Foley, J.A.; Ramankutty, N.; Brauman, K.A.; Cassidy, E.S.; Gerber, J.S.; Johnston, M.; Mueller, N.D.; O'Connell, C.; Ray, D.K.; West, P.C.; et al. Solutions for a cultivated planet. *Nature* **2011**, *478*, 337–342. [CrossRef] [PubMed]
2. Xu, J.; Liu, C.; Hsu, P.C.; Zhao, J.; Wu, T.; Tang, J.; Liu, K.; Cui, Y. Remediation of heavy metal contaminated soil by asymmetrical alternating current electrochemistry. *Nat. Commun.* **2019**, *10*, 1–8. [CrossRef] [PubMed]
3. Bo, S.; Luo, J.; An, Q.; Xiao, Z.; Wang, H.; Cai, W.; Zhai, S.; Li, Z. Efficiently se-lective adsorption of Pb(II) with functionalized alginate-based adsorbent inbatch/column systems: Mechanism and application simulation. *J. Clean. Prod.* **2020**, *250*, 119585. [CrossRef]
4. Adriano, D.C.; Wenzel, W.W.; Vangronsveld, J.; Bolan, N.S. Role of assisted natural remediation in environmental cleanup. *Geoderma* **2004**, *112*, 121–142. [CrossRef]
5. Amundson, R.; Berhe, A.A.; Hopmans, J.W.; Olson, C.; Sztein, A.E.; Sparks, D.L. Soil and human security in the 21st century. *Science* **2015**, *348*, 1261071. [CrossRef] [PubMed]
6. Liu, L.; Li, W.; Song, W.; Guo, M. Remediation techniques for heavy metal-contaminated soils: Principles and applicability. *Sci. Total Environ.* **2018**, *633*, 206–219. [CrossRef] [PubMed]
7. Park, K.O. A Study on Soil Washing Efficiency of Copper(Cu) Contaminated Soil in Military Rifle Range Site. Master's Thesis, Kyungpook National University Graduate School of Industry, Daegu, Korea, 2013.
8. Chen, R.; Sherbinin, A.D.E.; Ye, C.; Shi, G. China's soil pollution: Farms on the frontline. *Science* **2014**, *344*, 691. [CrossRef] [PubMed]
9. Yang, H. China's soil plan needs strong support. *Nature* **2016**, *536*, 375. [CrossRef] [PubMed]
10. Liu, B.L.; Ai, S.W.; Zhang, W.Y.; Huang, D.J.; Zhang, Y.M. Assessment of the bioavailability, bioaccessibility and transfer of heavy metals in the soil-grain-human systems near a mining and smelting area in NW China. *Sci. Total Environ.* **2017**, *609*, 52–60. [CrossRef] [PubMed]
11. Fei, X.F.; Christakos, G.; Xiao, R.; Ren, Z.Q.; Liu, Y.; Lv, X.N. Improved heavy metal mapping and pollution source apportionment in Shanghai City soils using auxiliary information. *Sci. Total Environ.* **2019**, *661*, 168–177. [CrossRef] [PubMed]
12. Rui, D.; Wu, Z.; Ji, M.; Liu, J.; Wang, S.; Ito, Y. Remediation of Cd- and Pbcontaminated clay soils through combined freeze-thaw and soil washing. *J. Hazard. Mater.* **2019**, *369*, 87–95. [CrossRef] [PubMed]
13. Tao, Y.; Huang, H.; Zhang, H. Remediation of Cu-phenanthrene co-contaminated soil by soil washing and subsequent photoelectrochemical process in presence of persulfate. *J. Hazard. Mater.* **2020**, *400*, 123111. [CrossRef] [PubMed]
14. Trellu, C.; Pechaud, Y.; Oturan, N.; Mousset, E.; van Hullebusch, E.D.; Huguenot, D.; Oturan, M.A. Remediation of soils contaminated by hydrophobic organic compounds: How to recover extracting agents from soil washing solutions? *J. Hazard. Mater.* **2021**, *404*, 124137. [CrossRef] [PubMed]
15. Shin, E.C.; Lee, M.S.; Park, J.J. A Study on the Numerical Analysis for Soil Contamination Prediction in Incheon Area. *J. Korea Geosynth. Soc.* **2012**, *11*, 21–30. [CrossRef]
16. Song, B.; Zeng, G.; Gong, J.; Liang, J.; Xu, P.; Liu, Z.; Zhang, Y.; Zhang, C.; Cheng, M.; Liu, Y.; et al. Evaluation methods for assessing effectiveness of in situ remediation of soil and sediment contaminated with organic pollutants and heavy metals. *Environ. Int.* **2017**, *105*, 43–55. [CrossRef] [PubMed]
17. Kjeldsen, P.; Barlaz, M.A.; Rooker, A.P.; Baun, A.; Ledin, A.; Christensen, T.H. Present and long-term composition of MSW landfill leachate: A review. *Crit. Rev. Environ. Sci. Technol.* **2002**, *32*, 297–336. [CrossRef]
18. Han, W.J.; Fu, F.L.; Cheng, Z.H.; Tang, B.; Wu, S.J. Studies on the optimum conditions using acid-washed zero-valent iron/aluminum mixtures in permeable reactive barriers for the removal of different heavy metal ions from wastewater. *J. Hazard. Mater.* **2016**, *302*, 437–446. [CrossRef] [PubMed]
19. Han, Z.Y.; Ma, H.N.; Shi, G.Z.; He, L.; Wei, L.Y.; Shi, Q.Q. A review of groundwater contamination near municipal solid waste landfill sites in China. *Sci. Total Environ.* **2016**, *569*, 1255–1264. [CrossRef] [PubMed]
20. You, J.H. Bio-Molecula Separation Using Nano Porous Membrane On-Chip. Master's Thesis, Seoul National University Graduate School, Seoul, Korea, 2009.
21. Kim, N.H. Fabrication and characterization of porous membrane foe high precision gas filter by in-situ reduction/sintering process. Ph.D. Thesis, Hanyang University, Seoul, Korea, 2010.
22. Ministry of Environment. *White Paper of Environment 2018*; Ministry of Environment: Sejong-si, Korea, 2018.

MDPI
St. Alban-Anlage 66
4052 Basel
Switzerland
Tel. +41 61 683 77 34
Fax +41 61 302 89 18
www.mdpi.com

Applied Sciences Editorial Office
E-mail: applsci@mdpi.com
www.mdpi.com/journal/applsci